Mohamed Tarek Khadir

Artificial Neural Networks in Food Processing

Also of Interest

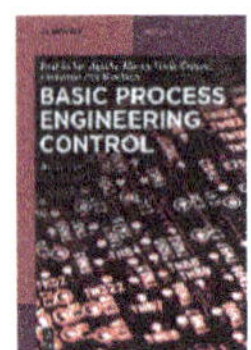

Basic Process Engineering Control
Paul Serban Agachi, Mircea Vasile Cristea and Emmanuel Pax Makhura (Eds.), 2020
ISBN 978-3-11-064789-1, e-ISBN 978-3-11-064793-8

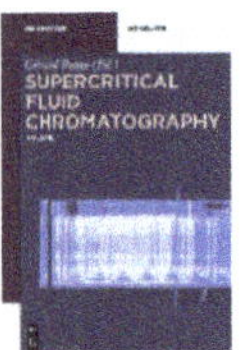

Supercritical Fluid Chromatography, Volume 1
Gérard Rossé (Ed.), 2019
ISBN 978-3-11-050075-2, e-ISBN 978-3-11-050077-6

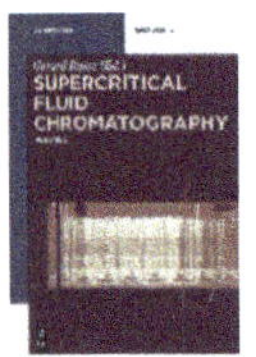

Supercritical Fluid Chromatography, Volume 2
Gérard Rossé (Ed.), 2019
ISBN 978-3-11-061893-8, e-ISBN 978-3-11-061898-3

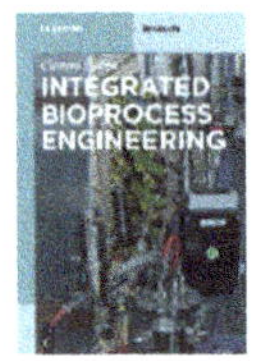

Integrated Bioprocess Engineering
Clemens Posten, 2018
ISBN 978-3-11-031538-7, e-ISBN 978-3-11-031539-4

Mohamed Tarek Khadir

Artificial Neural Networks in Food Processing

Modeling and Predictive Control

DE GRUYTER

Author
Prof. Mohamed Tarek Khadir
University Badji Mokhtar of Annaba BP12
Department of Computer Science
23000 Annaba
Algeria
khadir@labged.net

ISBN 978-3-11-064594-1
e-ISBN (PDF) 978-3-11-064605-4
e-ISBN (EPUB) 978-3-11-064613-9

Library of Congress Control Number: 2020945602

Bibliographic information published by the Deutsche Nationalbibliothek
The Deutsche Nationalbibliothek lists this publication in the Deutsche Nationalbibliografie; detailed bibliographic data are available on the Internet at http://dnb.dnb.de.

Cover image: DKosig / Gettyimages
Typesetting: VTeX UAB, Lithuania
Printing and binding: CPI books GmbH, Leck

www.degruyter.com

To Liam and Dana, with all my love. Thank you both for all the daily joy you give me watching you grow

Acknowledgement

Warmest thanks to Pr. John Ringwood, not only for having been my surpriser during my Doctoral years, but for being the first one to introduce me through his lectures to Artificial Neural Networks, in 1997. Parts of the introductory material here, are largely inspired by his lectures.

Finally, deepest thanks to all my Master and Ph.D. students that helped with this project.

https://doi.org/10.1515/9783110646054-201

Contents

Introduction

Humanity is reaching an important crossroad, where it will have to make a conceptual choice concerning its way of life, consumption, and relationship with nature, where mankind will have to reconsider his position in the ecosystem, switching from viewing nature as a reservoir to serve his needs, to be fully a part of that ecosystem. The Covid-19 pandemic, is in that sense, an electroshock and a serious reminder of the changes that has to be made. This change will necessarily involve the transition to a new economy, no longer based on mass production using in most cases polluting and harmful fossil energy resources.

The transition to a new economy, based mostly on renewable and green energy, may be achieved by a model greatly inspired by the most perfectly known, renewable ecosystem: "nature." The basis of this economy can only be constructed leaning on new technologies, not only for increasing production, but also technologies which allow us to push thinking, design, and computing to a higher level. The new paradigms of Artificial Intelligence (AI) are already starting to contribute greatly in those changes.

Information, as well as collaborative computational capabilities sharing, will be the basis of the new industrial revolution, which will have to be centred on people and their well-being in their natural protected environment. AI is the natural culmination of advances in information theory, in microelectronics, and the ever-increasing need to deal with overwhelming data and increasingly complex problems. The functioning of its different paradigms are mainly bio-inspired (e. g., Artificial Neural Networks, Genetic Algorithms, Swarm optimisation, etc.) with a reasoning philosophy being, as close as possible to human reasoning, acquiring with time and research, great generalisation and adaptation power. Connectionist approaches stemming from biomimetics are particularly important and among the most used branches of AI.

This book is doubly relevant in that sense, as it deals with Artificial Neural Networks (ANN), their foundations, history as well as the description of the most used types with relevant applications in food industry; vital sector for our food security, economy, and well-being. The book is both educational and academic, accessible to Master and PhD students wanting to familiarise themselves with ANN and may guide them step-by-step from biological inspiration and simple formal neuron to network construction, training, and validation in order to better understand and master most used types of ANN, namely: the MultiLayer Perceptron (MLP), Radial Base Networks (RBFs), and Kohonen maps or Self-Organizing Maps (SOM). Wider consideration is given to Deep Neural Networks (DNNs), which imposed themselves as one of the best solutions to solve complex high-dimensional problems governed by an important number of data scattered over a large input/output space.

An important part of the book deals with DNNs basics, explaining and detailing the most famous and used topologies. In this perspective, the main topologies such as Convolutional neural networks (CNN), Stack Denoising Auto-encoders (SDAE), and

https://doi.org/10.1515/9783110646054-202

Bolzman Machine (BM) are detailed and their learning algorithms formulated mathematically. In addition to the theoretical aspect of ANNs, in their simplest and most complex formulations (DNNs and Deep Learning (DL)), the initiated reader interested in the food industry will find an overview of practically everything that has been published merging ANNs and food applications. The survey covers more than 250 applications giving slight details of each one and presenting them in summary tables to promote more effective search and indexation. The use of DNNs in the food industry, although minimal and recent, is also covered and all applications are classified, indexed, and referenced. The reader will then be able to appreciate the use of a given type of networks, topology or paradigm for a specific given application, procedure, and product.

In the listed ANN applications, a wide spectrum of usage, ranging from classification and recognition of food products to modelling and forecasting of productive behaviour, storage, and others are covered. However, the author wanted to highlight the use of ANNs in the regulation and control procedures of industrial food processes, more particularly the use of ANN in the control and regulation procedures based on Model Predictive Control (MPC).

Indeed, the fact that MPC has proven, since the 1970s, its effectiveness for the control and regulation of industrial processes over other types of advanced and other model-based control strategies, especially in the refining and petrochemical industry, making it conquer more markets as an essential and interesting control strategy prospect. In addition to its efficiency mainly due to the forecast of the process output, using an internal embedded model, the receding horizon control approach gives the regulator an ability to reduce the variances of the control and output variables within the limits of existing constraints. Initially, the internal model is based on simple linear mathematical models in order to easily derive an analytical control law. The emergence of ANNs, Nonlinear Model Predictive Control (NMPC) structures based on these types of models have emerged. All nonlinear ANN based MPC applications in the food industry are listed and referenced in this book. Finally, and in order to demonstrate the implementation steps for all the ANNs models and ANN based approaches described, an application from different applications and processes in food industry is presented for every presented ANN paradigm, with all the results and comments issued from real data and processes experimentations.

Finally, the reader will find a case study for every ANN topology or paradigm presented in the book, detailing the conceptual, training, and validation stages of the procedures. For instance, the usage of a MLP classifier for wine producer is given in Section 3.10, where 13 wine characteristics are used as inputs to correctly classify a wine sample into three different producer classes. In the same way, a RBF is used to classify oil origin (from four producing countries). The RBF takes as inputs, this time, more than 500 parameters for a single input oil sample issued from a spectral analysis in Section 4.4. A clustering case study is given in Section 5.3 using Kohonen maps in order to find clusters of Substrate Methane (biogas) Production issued from food waste.

In this case, no output is specified and the inputs are the methane production time series for each food and agricultural waste. Advanced ANNs in the form of DNNs and more precisely Convolutional Neural Networks (CNNs) and Stacked Denoising Auto-Encoders (SDAEs) are used to classify two classes of fruit images in Section 6.8. Finally, a detailed case study of Model Predictive control of pasteurisation temperature in a plate heat exchanger, using an ANN internal model is giving in Section 7.5. All obtained results are based on validated real life processes data. This strengthens all drawn conclusions and will, hopefully, help the reader consider and develop an ANN based solution in the food industry.

All the codes for the presented examples in Chapters 3, 4, 5, 6 and 7 for applications using respectively MLPs, RBFs, Kohonen Maps, Deep ANNs and NMPC can be found at: https://www.degruyter.com/view/title/551147.

1 Biological inspiration and single artificial neurons

1.1 Introduction and biological inspiration

Neural networks are used to solve complex problems in nonlinear optimisation as well as classification problems. The techniques used are based on the neuron operations functioning in a human or animal brain, hence the biological inspiration.

A single brain neuron can be considered as an automaton (see Figure 1.1) comprising:

The central cell (or soma) containing the nucleus: it is a few microns in diameter performs the biochemical transformations necessary for the synthesis of enzymes and other molecules that ensures the life of the neuron.

Dendrites receiving signals from other neurons via synapses: These are a few tenths of a micron in diameter and a length of tens of microns.

The axon that carries information out of the cell to other neurons: The axon is usually longer than the dendrites, it communicates with other neurons by its end.

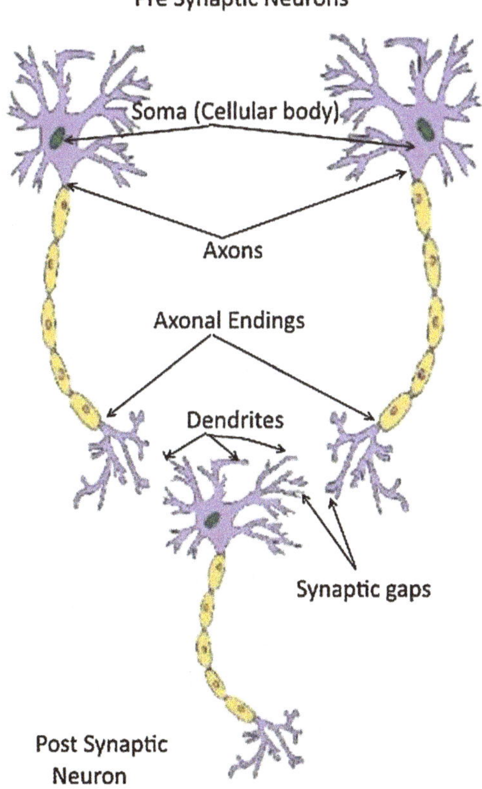

Figure 1.1: Biological neuron.

https://doi.org/10.1515/9783110646054-001

The connections between two neurons are made in places called synapses where they are separated by a small synaptic space of the order of a hundredth of microns. The set of input signals (excited and inhibited) is averaged. If this average is large enough over a very short period of time, the cell delivers a signal electrical to the following cells via his axon.
Delivered impulses (or action potential) are of "all-or-nothing type", i. e., the neurons communicate with each other in binary language.

The main characteristics of a biological neuron can be listed in what follows:
- A human brain contains on average between 10^{10} and 10^{12} neurons. Neurons are interconnected to each other in a very complex spatial form to form the nervous system.
- The core (or Soma) has diameters of a few microns. Dendrites are rarely longer than 10 microns and receive signals entries.
- The axon forms the exit of the neuron, it has a length between 1 mm and can reach 1 m long. It can connect the neurons to one or more other neurons, where a neuron can be connected to a hundred or even thousands of other neurons.
- The signals are electrochemical in nature. The propagation speed varies from 5 to 125 m/s A delay of 1 ms is necessary for the signal to cross the synapses.
- It is observed that through intense brain activity metabolic development occurs at the synaptic level. This leads to an enlargement of the synaptic surface and, therefore, an increase in the weight of a particular input.
- If the sum of the signals received by a neuron exceeds a certain limit, the latter is activated. Neurons can be active on a wide frequency range, but always at the same amplitude.
- After being activated an axon is out of service for a period of 10 m/s. The information is frequency coded for the transmitted signals.

A mathematical interpretation of the biological neuron can then be formulated. The neuron receives a variable number of inputs from upstream neurons. At each of these inputs is associated a weight w_i representative of the strength of the connection. Each elementary processor has a single output, which then branches out to power a variable number of downstream neurons. The biological inspiration is obvious where the soma is given by the nonlinear transfer function receiving input signals represented by dendrites and delivering the response via the output representing the axon and the synaptic space or synaptic weight affecting the flow of the information is represented by an evolving numerical weight, to be determined and fixed after learning. The final formal neuron can be given in Figure 1.2, and is known as McCulloch–Pitt neuron.

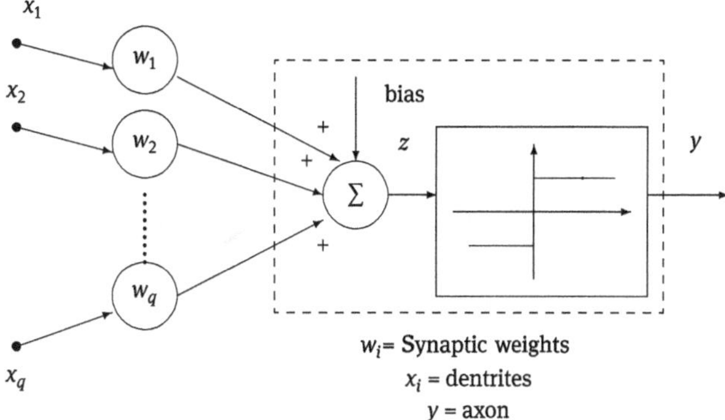

Figure 1.2: Artificial neuron.

1.2 Brief history of ANN development

ORIGINS

- 1890: W. James, famous American psychologist, introduces the concept of associative memory, and proposes what will become a functioning learning law for neurone training, later known as Hebb's Learning Rule in 1949 [1].
- 1943: J. Mc.Culloch and W. Pitts are the first to show that a simple formal neuron can perform logical, arithmetic, and complex derivations [2].
- 1949: D. Hebb, American physiologist, explains conditioning in animal behaviours by the neurons themselves. Therefore, a Pavlovian type of conditioning such as feeding all day at the same time a dog leads in this animal to the secretion of saliva at that particular time even when there is no food. The proposed modification law of the connection properties between neurons partly explains this type of experimental results.
- 1957: F. Rososenblatt develops the model of the perceptron. He builds the first neuro-computer based on this model and applies it to the field of pattern recognition [3].
- 1960: B. Widrow, a control engineer, develops the Adaline model (Adaptative Linear Element) in its structure, the model resembles the perceptron. However, the learning law is different [5, 12, 14].

THE QUIET YEARS

- 1969: M. Minsky and S. Papert published a paper that highlights the theoretical limitations of the perceptron. Limitation concerning in particular the impossibility of treating by this model, nonlinearly separable problems [6].
- 1967–1982: All research continues, but focusing solemnly on various application areas. Big names working during this period include S. Grossberg, T. Kohonen. Albus developed its adaptive "Cerebral Model Articulation Controller," which is

summarised in a sort of correspondence table, imitating the perception of human memory [7].

INTEREST REVIVAL

- 1982: J. J. Hopfield, a recognised physicist, to whom we owe the revival of interest for ANNs with his ANN Hopefield model. Noting that at this date the AI is the object of some disillusionment, it did not answer all expectations and even faced serious limitations. Even though the limitations of the perceptron put forward by Minsky and Papert are not yet lifted by the Hopefield model, this one may find a solution when a single perceptron cannot [8].
- 1983: The Boltzmann machine is the first known model to satisfactorily treat Minsky and Papert identified limitations of the perceptron [15].
- 1985: Gradient backpropagation is developed by three groups of independent researchers. Nowadays, the multi-layer Perceptron and the gradient backpropagation learning rule, remains the most studied and the most productive model in terms of number of applications [17]. An architecture similar to the MLP is the Probabilistic Neural Network (PNN), which divides the latter by its exponential activation function and the forms of its connections [18].
- 1987: The confirmation of Hopfield networks as a reference for associative memory network problem. Later, Kosko will extend this concept to develop his "Bidirectional Associative Memory" (BAM) [16].
- 1988: Broomhead and Lowe (1988), inspired by radial functions, developed Radial Basis Function (RBF) networks [19]. They opened the way to Chen and Billings (1992) and their functional Link Network (FLN) where a nonlinear transformation of the inputs allows a reduction of the computing power as well as faster convergence [20].
- 1997: Silicon implementation of ANNs (Very large-scale integration (VLSI) of ANNs) which raises off a lot of speed limitations and offered possibilities not yet envisaged until then.
- 2000–2010: The growing of machine learning and the establishment of different techniques and learning paradigms as a replacement for traditional algebraic and "frozen" solutions. The time lap has seen the emergence and the spread usage of popular supervised and unsupervised training algorithms such as: Competition learning, reinforcement learning, and the improvement of the backpropagation algorithm.
- 2010 to date: Since 2010, and with access to Big Data and accessibility of extremely powerful GPUs, it was possible to consider increasingly dense network topologies and termed Deep layers Neural Networks (DNNs) became possible to train. Yann LeCun, a machine learning specialist, is one of the fathers of Deep Learning; a method to which he has devoted himself for 30 years, despite the skepticism he

initially encounters in the scientific community. Since 2012, large artificial intelligence projects have been set up by IBM (Watson), Google (DeepMind/AlphaGo) and Facebook (DeepFace); see Chapter 6.

1.3 Artificial neural network types

ANN can be found in many forms for different purposes, implementing supervised or unsupervised learning approaches. They can be distinguished and categorised by the three following characteristics:

- Network topology: Number of layers, number of neurons, connection types (feedforward, recurrent), network shape, etc.
- Activation function: Sigmoidal for a MLP, radial basis function for RBFs, etc.
- Training algorithm: Backpropagation for MLPs, competition for Kohonen maps, LMS rule for RBFs, etc.

Despite their variety, one may cluster ANNs in five main types if deep neural network is considered. Note that the most represented network types will be later described in detail.

1. **Multi-layered Perceptron (MLP):** This is the most used artificial neural network and the most reported in literature until now. MLPs have proven their wide efficiency when it comes to supervised learning problems, either for classification or function approximation. A wide range of libraries in different computer languages are freely available. Chapter 3 is devoted to explaining the architecture and training of MLPs, supported by both academic and real-life examples.
2. **Radial basis functions (RBF):** This is a single hidden layer network composed of Radial Basis Function neurons. RBF are trained in two phases. The first one will define the number, centres, and width of the neurons while the second computes, after training, the weights of the output layer [13]. Chapter 4 details RBF networks, explains their construction and training, the whole supported by academic and real life examples.
3. **Hoppefield networks:** In these networks, each input vector is associated with an output vector instead of a scalar. The output vector can either represent a form identified by the network or classifies the input vector into one of several categories, each represented by an output neuron. This is the case for all auto-associative memory networks and use Hebbian-inspired learning rules or delta rules [12]. Hoppefield networks may be considered as energy minimizing networks: they are composed of fully interconnected neurons; they do not have an "entry" and an "exit" but they evolve from an initial state. In general, a Hopfield network is defined by n neurons and by a transfer symmetrical matrix, W [8].
4. **Kohonen self-organising feature maps (SOFM) or Self Organising Maps (SOM):** They are the simplified model of the notion of characteristics related to a

given region of the brain, where each part of the map is associated, after training, to a particular cluster or characteristics. Competition unsupervised learning is mainly used for training the Kohonen map. The main goal for their usage is data clustering and dimension reduction in data. They can be used in many dimensions, but 2 or 3 dimension maps are preferred for visualisation purposes [21]. A network similar to SOFM is the CelNN (Cellular Neural Network), which is a kind array with dynamic systems, called cells, that are connected locally [22]. Each cell interacts only with adjacent cells.
Chapter 5 is devoted to detailing SOMs, explaining their functioning and the tracing procedure used for weight's update. The design of SOM for clustering real life data in the domain of food industry is shown for clustering food waste for biogas production.
5. **Deep neural networks:** They can be classified as large MLPs, with preliminary layers usually devoted to characteristics extraction, with the main difference lying in the large number of hidden layers and the learning algorithms used. A substantial numbers of deep architectures are discussed in Chapter 6, where most famous DNN architectures are explained to the reader, their training mechanisms detailed and supported by real life examples for fruit images classification.

The above mentioned types of ANNs have been increasingly used in the industry since the late 1980s. In 2003, Magali et al. presented a comprehensive review of industrial applications of artificial neural networks from 1990 [9], where most popular ANN architectures along with training approaches and algorithms helped ANN users in problem solving.

Magali et al. stressed that although ANNs have been used at the time, for more than 50 years, their effectiveness and widespread usage started in the 1990s. It was proven then that they could provide an improvement or at least a complement to previously used modelling and decision-making techniques.

The ANN developed applications helped improve or find suitable solutions to industrial problems, thus democratizing the concept and making it a valuable tool for industrial processes applications.

Table 1.1 summarises the most common ANN architectures derived from the four main types described earlier when used for optimisation, associative memory, pattern recognition, function approximation, modelling and control, image processing, and classification purposes. A more specific review concerning solemnly food industry will be presented in Chapter 2.

More recently and in order to evaluate the progress of ANN usage in the 21st century, Abiodun et al. [10] presented a survey of ANN applications in the real industrial processes. A taxonomy of ANNs is provided and furnished the reader with knowledge of current and emerging trends in ANN applications research and area of focus. The study, similar to [9] earlier, covers many applications using different ANN paradigms and architectures in various disciplines, which include computing, science, medicine,

Table 1.1: ANN Applications per fields.

Functional Characteristics	Structure
Pattern Recognition	MLP, Hopefield, Kohonen, PNN
Associative Memory	Hopefield, Recurent MLP, Kohonen
Function Approximation	MLP, RBF
Optimisation	Hopefield, ART, CNN
Classification and Clustering	MLP, Kohonen, RBF, ART, PNN
Modelling and Control	MLP, Reccurent MLP, FLN
Image Processing	CNN, Hopefield
Classification (including Clustering)	MLP, Kohonen, RBF, ART, PNN

engineering, agriculture, mining, environmental, technology, climate, business, arts, and nanotechnology, etc. The study assesses ANN contributions, compared performances, and critique methods. However, in addition to [9], deep learning is introduced and assessed in real life application until 2018. The results of the study in terms of number of ANN application found is depicted in Table 1.2. For details on used ANN architectures, the reader is directed to the paper [10]. One of the main conclusions of the study is that neural-network models such as feedforward and feedback propagation artificial neural networks are performing better for human related problems, such as recognition, aided decision, etc.

Table 1.2: ANN and deep ANN application numbers per field (2018) [10].

Application Field	Prediction	Pattern recognition	Classification	Total
Security	20	18	2	40
Science	25	25	2	52
Engineering	22	7	2	31
Medical science	10	5	2	17
Agriculture	3	3	2	7
Finance	10	15	2	27
Bank	5	15	2	22
Weather and climate	2	15	2	19
Education	30	15	2	47
Environmental	10	15	2	27
Energy	5	15	2	22
Mining	2	15	2	19
Policy	2	2	2	6
Insurance	5	4	2	11
Marketing	5	5	2	12
Management	40	2	2	44
Manufacturing	12	15	5	32
Other fields	52	11	10	71

1.4 Simple formal neurons

1.4.1 Mc Culloch–Pitt neuron

Mc Culloch–Pitt neuron 1943, is the first-ever mathematically formalised neuron and is shown in Figure 1.3. It is characterised by a limitation transfer function of 0/1 type ($s = 1$ or 0) [2].

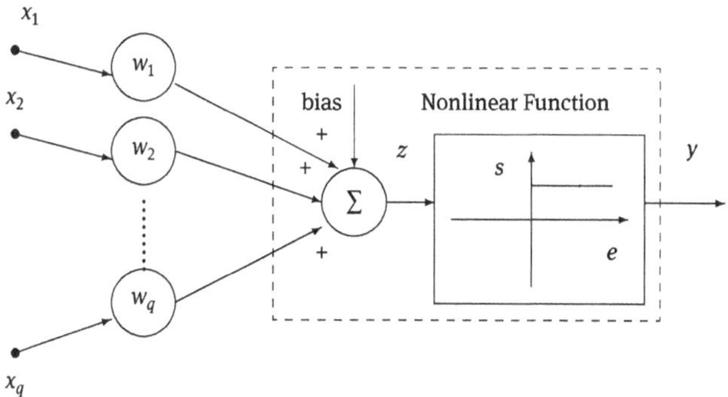

Figure 1.3: Mc Culloch–Pitt neuron.

1.4.2 The perceptron

The perceptron was designed by Rosenblatt in 1957 and was a prelude for the perceptron learning rule, Figure 1.4, and is characterised by [4]:

- a continuous transfer function of a sigmoidal form: $y = \frac{2}{1+e^{-\beta z}} - 1$;
- a squashing transfer function limited by: $-1 < y < 1$;
- the transfer function is derivable. This is very important in order to implement further learning rules (i. e., the Delta learning rule).

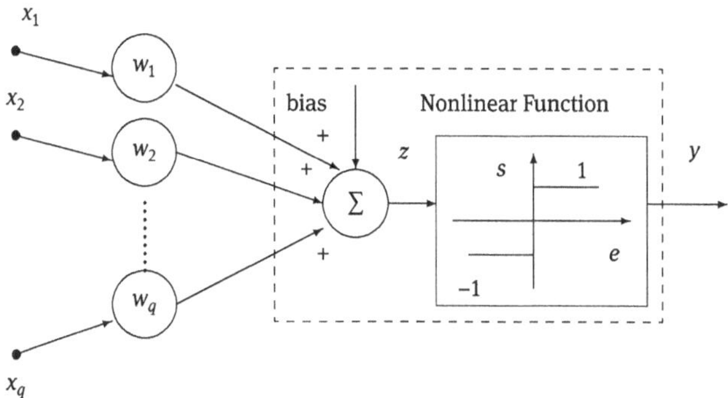

Figure 1.4: The perceptron.

1.5 Hebb learning rule (1949)

The Hebb learning rule is probably the first 'Pavlovian" inspired mathematical learning rule for a single neuron. The learning type used is unsupervised and is depicted in Figure 1.5 as opposed to the supervised type [1].

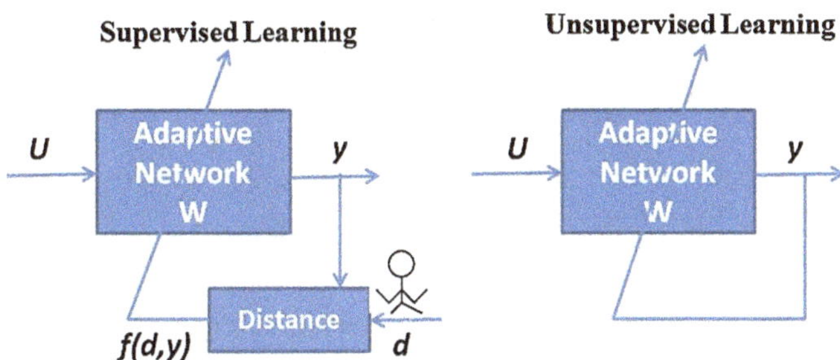

Figure 1.5: Learning approaches for adaptive networks.

In the unsupervised approach, no observers are present, and presented data or examples are not organised in classes nor have known outputs. In other words:
- No class and no target known.
- The general goal is then to discover classes or groups of similar examples.

The rule is inspired from the mechanism of biological neuron where: when the presynaptic axon of neuron A is close to excite a post synaptic neuron B (see Figure 1.1); this will induce A to fire. A metabolic change or evolution, then occurs in neuron B. In other words, the efficiency of A firing to B is increased.

Algorithm.

$$y = f(W^T X) \tag{1.1}$$

$$\Delta W = \mu y X = \mu f(W^T X)X \tag{1.2}$$

For neuron i and input j, we have

$$\Delta w_{ij} = \mu y X = \mu f(w_i^T X_j)X_j \tag{1.3}$$

with:
- W_0 is randomly initialised.
- The training process is purely feedforward and nonsupervised.
- The weights W_i, will be strongly influenced by frequent similar inputs.
- A nonconstrained evolution of W_i.
- The learning rule is intuitive and purely based on correlation.

1.5.1 The Hebb learning rule applied to a sign-type neuron

Let the sign-type neuron presented in Figure 1.6, and is a version of the Perceptron neuron with $\beta \to \infty \Rightarrow f(W^T X) = \operatorname{sgn}(W^T X)$.

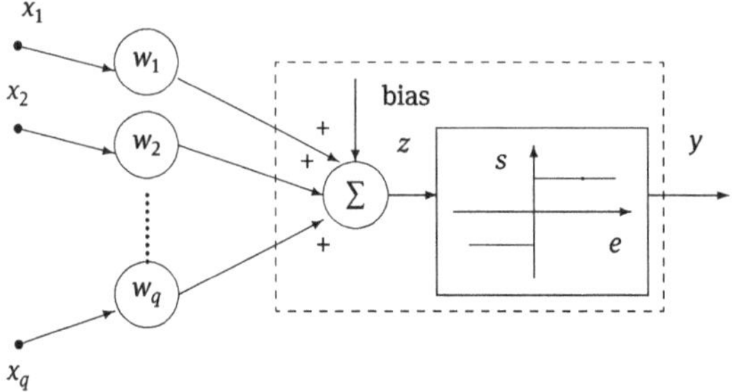

Figure 1.6: Sign-type neuron.

Let:

$$W_0 = \begin{bmatrix} 1.0 \\ -1.0 \\ 0.0 \\ 0.5 \end{bmatrix} \quad \text{and} \quad X_0 = \begin{bmatrix} 1.0 \\ -2.0 \\ 1.5 \\ 0.0 \end{bmatrix}, \quad X_1 = \begin{bmatrix} 1.0 \\ -0.5 \\ -2.0 \\ -1.5 \end{bmatrix}, \quad X_2 = \begin{bmatrix} 0.0 \\ 1.0 \\ -1.0 \\ 1.5 \end{bmatrix}$$

If for simplification $\mu = 1$, we obtain

$$z_0 = W_0^T X_0 = [1.0 \ -1.0 \ 0.0 \ 0.5] \begin{bmatrix} 1.0 \\ -2.0 \\ 1.5 \\ 0.0 \end{bmatrix} = 3$$

The weight update ($W_{k+1} = W_k + \Delta W$) is performed as shown below:

$$W_{k+1} = W_k + \operatorname{sgn}(z_k) X_k \tag{1.4}$$

where

$$W_1 = \begin{bmatrix} 1.0 \\ -1.0 \\ 0.0 \\ 0.5 \end{bmatrix} + 1. \begin{bmatrix} 1.0 \\ -2.0 \\ 1.5 \\ 0.0 \end{bmatrix} = \begin{bmatrix} 2 \\ -3.0 \\ 1.5 \\ 0.5 \end{bmatrix}$$

For the two next iterations:

$$z_1 = -0.25, \quad W_2 = W_1 + \operatorname{sgn}(z_1) X_1 = W_1 - X_1 = \begin{bmatrix} 1.0 \\ -2.5 \\ -0.5 \\ -1.0 \end{bmatrix}$$

$$z_2 = -3.0, \quad W_3 = W_2 + \mathrm{sgn}(z_2)X_2 = W_2 - X_2 = \begin{bmatrix} 1.0 \\ -3.5 \\ -1.5 \\ 1 \end{bmatrix}$$

In conclusion, the Hebb learning rule used with a sign-type neuron, comes to add or subtract the input vector from the weights, of course, if $\mu = 1$.

1.5.2 Hebb learning rule used for a Perceptron type neuron

Let us consider the Perceptron type neuron, Figure 1.4, with $\beta = 1 \Rightarrow f(W^T X) = y = \frac{2}{1+e^{-z}} - 1$.

Let:

$$W_0 = \begin{bmatrix} 1.0 \\ -1.0 \\ 0.0 \\ 0.5 \end{bmatrix} \quad \text{and} \quad X_0 = \begin{bmatrix} 1.0 \\ -2.0 \\ 1.5 \\ 0.0 \end{bmatrix}, \quad X_1 = \begin{bmatrix} 1.0 \\ -0.5 \\ -2.0 \\ -1.5 \end{bmatrix}, \quad X_2 = \begin{bmatrix} 0.0 \\ 1.0 \\ -1.5 \\ 1.5 \end{bmatrix}$$

If we consider, for simplification $\mu = 1$, we obtain

$$z_0 = W_0^T X_0 = [1.0 \ -1.0 \ 0.0 \ 0.5] \begin{bmatrix} 1.0 \\ -2.0 \\ 1.5 \\ 0.0 \end{bmatrix} = 3$$

The weights update is performed as usual:

$$W_{k+1} = W_k + f(z_k)X_k, \quad y = f(z) = y = \frac{2}{1+e^{-\beta z}} - 1 \tag{1.5}$$

where

$$W_1 = \begin{bmatrix} 1.0 \\ -1.0 \\ 0.0 \\ 0.5 \end{bmatrix} + 0.905 \begin{bmatrix} 1.0 \\ -2.0 \\ 1.5 \\ 0.0 \end{bmatrix} = \begin{bmatrix} 1.905 \\ -2.810 \\ 1.357 \\ 0.500 \end{bmatrix}$$

for the next two iterations:

$$y_1 = -0.077, \quad W_2 = W_1 + 0.077X_1 = \begin{bmatrix} 1.982 \\ -2.8485 \\ 1.2035 \\ 0.3845 \end{bmatrix}$$

$$y_2 = -0.932, \quad W_3 = W_2 + 0.932x_2 = \begin{bmatrix} 1.9820 \\ -1.9165 \\ -0.1945 \\ 1.7825 \end{bmatrix}$$

The Hebb learning rule used with a continuous neuron of Perceptron type, comes to add or subtract a fraction (determined by β) of the inputs and the weight vector.

1.6 Perceptron learning rule

The perceptron learning rule was introduced by Rosenblatt, for his Perceptron neuron in 1958. The learning rule as opposed to Hebb rule is supervised [3]. In supervised learning the consequences (output, classes, etc.) are known for each example, and an observer is then used to assess at each iteration the difference between the real output and the one given by the perceptron output. The learning rule is characterised by:
- The rule can be applied to sign-type neurons.
- The training signal is the difference between the actual neuron output and the desired response; see Figure 1.5.
- The presence of a known desired output implies supervised learning.

The Perceptron neuron with training signal d is shown in Figure 1.7.

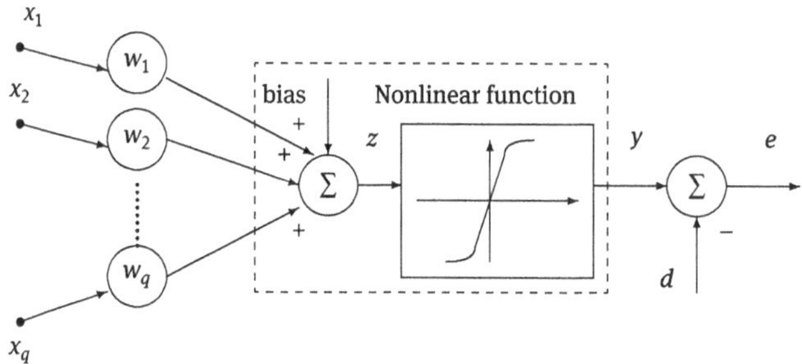

Figure 1.7: Neuron type used with the Perceptron learning rule.

Algorithm.

$$e = d - y, \quad y = \text{sgn}(W^T X) \tag{1.6}$$

$$\Delta W = \mu e X = \mu[d - \text{sgn}(W^T X)]X \tag{1.7}$$

considering $d = 1$ or -1,

$$\Delta W = \pm 2\mu X \tag{1.8}$$

Example. Let:

$$W_0 = \begin{bmatrix} 1.0 \\ -1.0 \\ 0.0 \\ 0.5 \end{bmatrix}, \quad d_0 = -1.0, \quad d_1 = -1, \quad d_2 = 1,$$

$$\text{and} \quad X_0 = \begin{bmatrix} 1.0 \\ -2.0 \\ 0.0 \\ -1.0 \end{bmatrix}, \quad X_1 = \begin{bmatrix} 0.0 \\ 1.5 \\ -0.0 \\ -1.0 \end{bmatrix}, \quad X_2 = \begin{bmatrix} -1.0 \\ 1.0 \\ 0.5 \\ -1.0 \end{bmatrix}$$

$$z_0 = W_0^T X_0 = [1.0 \ -1.0 \ 0.0 \ 0.5] \begin{bmatrix} 1.0 \\ -2.0 \\ 0.0 \\ -1.0 \end{bmatrix} = 2.5$$

Note that $\text{sgn}(2.5) \neq d_0$. If $\mu = 0.1$, we have

$$W_1 = W_0 + 0.1(-1 - 1)X_0$$

$$W_1 = \begin{bmatrix} 1.0 \\ -1.0 \\ 0.0 \\ 0.5 \end{bmatrix} - 0.2 \begin{bmatrix} 1.0 \\ -2.0 \\ 1.5 \\ -1.0 \end{bmatrix} = \begin{bmatrix} 0.8 \\ -0.6 \\ -0.3 \\ 0.7 \end{bmatrix}$$

$$z_1 = W_1^T X_1 = [0.0 \ 1.5 \ 0 \ -1.0] \begin{bmatrix} 0.8 \\ -0.6 \\ -0.3 \\ 0.7 \end{bmatrix} = -1.6$$

Considering, $\text{sgn}(-1.6) = d_1$, the weight update is not necessary.

$$z_2 = W_2^T X_2 = [1.0 \ -1.0 \ 0.5 \ -1.0] \begin{bmatrix} -1.0 \\ 1.0 \\ 0.5 \\ -1.0 \end{bmatrix} = -0.75$$

Note that $\text{sgn}(-0.75) \neq d_1$. The weight update is now necessary.

$$W_3 = W_2 + 0.1(1 + 1)X_2$$

$$W_3 = \begin{bmatrix} 0.8 \\ -0.6 \\ 0.0 \\ 0.7 \end{bmatrix} + 0.2 \begin{bmatrix} 1.0 \\ -1.0 \\ 0.5 \\ -1.0 \end{bmatrix} = \begin{bmatrix} 0.6 \\ -0.4 \\ 0.1 \\ 0.5 \end{bmatrix}$$

Remarks:
- There are no obvious justifications for the inclusion of the input X, in equation (1.7).
- The choice of the learning rate μ, is not specified.
- The Perceptron learning rule can be specified as below
  - With a McCulloch–Pitt neuron,
    $W_{k+1} = W_k + \mu X$ if $y = 0$ and $d = 1$
    $W_{k+1} = W_k - \mu X$ if $y = 1$ and $d = 0$
    $W_{k+1} = W_k$ if $y = d$
  - With a Perceptron of sign-type $\beta \to \infty$,
    $W_{k+1} = W_k + 2\mu X$ if $y = -1$ and $d = 1$
    $W_{k+1} = W_k - 2\mu X$ if $y = 1$ and $d = -1$
    $W_{k+1} = W_k$ if $y = d$
- It can be proven that the Perceptron learning rule and w, converge at each iteration.

1.7 Classification and linear separability

This section addresses the problem of classification and linear separability supported by a classifier given by a single neuron and its limitations as stated by Minski and Papert in [6].

1.7.1 Classification

Let us here describe formally what is "classification" in order then to propose a suitable or test any classifier. The goal of classification (or pattern classification) is to assign a physical object, event or phenomenon to a given identified class (or category). One may cite some classification examples:
- Boolean functions.
- Pixel patterns, e. g., seven segment display.
- Analog/digital transform.
- Associative memory, e. g., written character recognition.

1.7.2 Discriminant functions

Problem statement

Let us suppose that there is a pattern matrix of p dimension: $X_1, X_2, \dots, X_p$ and that the class of every pattern X_i is known among the n classes. The number or dimension of the input matrix of elements p has to be greater than the number of classes n. It is also reasonable to assume that n is more important than the number of categories R.

The belonging to a given category is defined or proven by a classifier based on the comparison of R discriminant functions $g_1(X), g_2(X), \ldots, g_R(X)$ computed for each input vector X_i. The element X_i belongs to a category i if and only if

$$g_i(X) > g_j(X) \quad \text{for } i,j = 1, 2, \ldots, R,\ i \neq j \tag{1.9}$$

The equation defining the decision frontier is

$$g_i(X) - g_j(X) = 0 \tag{1.10}$$

1.7.3 Linear separability

A neuron of sign type (threshold) implements the separation of an input vector into two classes, defined by

$$W^T X = -b \tag{1.11}$$

where b is the limit or bias. The frontier separating both classes is a hyperplane of n dimensions.

Remarks:

- The proportions of separable linear functions in the domain on n Boolean functions decrease exponentially with n.
- A network containing two layers of sign-type neurons, is able to model and simulate any boolean function by an implementation of sum/product (or product/sum) architecture type, Figure 1.8.

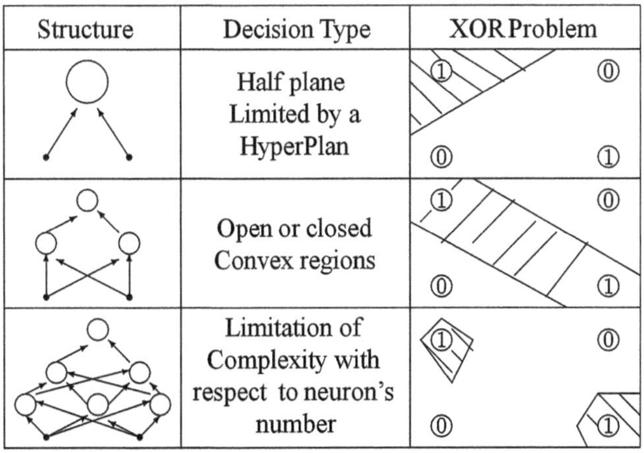

Structure	Decision Type	XOR Problem
	Half plane Limited by a HyperPlan	1 0 0 1
	Open or closed Convex regions	1 0 0 1
	Limitation of Complexity with respect to neuron's number	1 0 0 1

Figure 1.8: Classification of the XOR function using ANNs.

1.7.4 Example: Implementation of the AND function using a single neuron

The AND function here is implemented as a single neuron in Matlab using the functions *newp* to create a single or a neural network, a function *train* for training a single neuron of the Perceptron type.

The implementation details are given below:

- 2 inputs + 1 bias, with an initialisation $W_0 = [0.1\ \ 0.1]$ and $b(0) = 0.1$.
- A training of 8 epochs with a learning rate of 0.1.
- The decision surface is giving by: $x_2 = -\frac{w_1}{w_2}x_1 - \frac{b}{w_2}$.
- $P = \left[\begin{smallmatrix} 0 & 1 & 0 & 1 \\ 0 & 0 & 1 & 1 \end{smallmatrix}\right] = \left[\begin{smallmatrix} x_1 \\ x_2 \end{smallmatrix}\right]$, $T = [0\ \ 0\ \ 0\ \ 1]$.

It can be seen in Figure 1.9(c) that the initial defined goal (a training Mean Squared Error MSE = 0.0001) is obtained after 6 training epochs. The decision line evolution is shown in Figure 1.9(a), (b), (c), and (d), with the evolution of the separation line between 1 s and 0 s from epoch 1 to 15. It can be clearly seen that the separation is

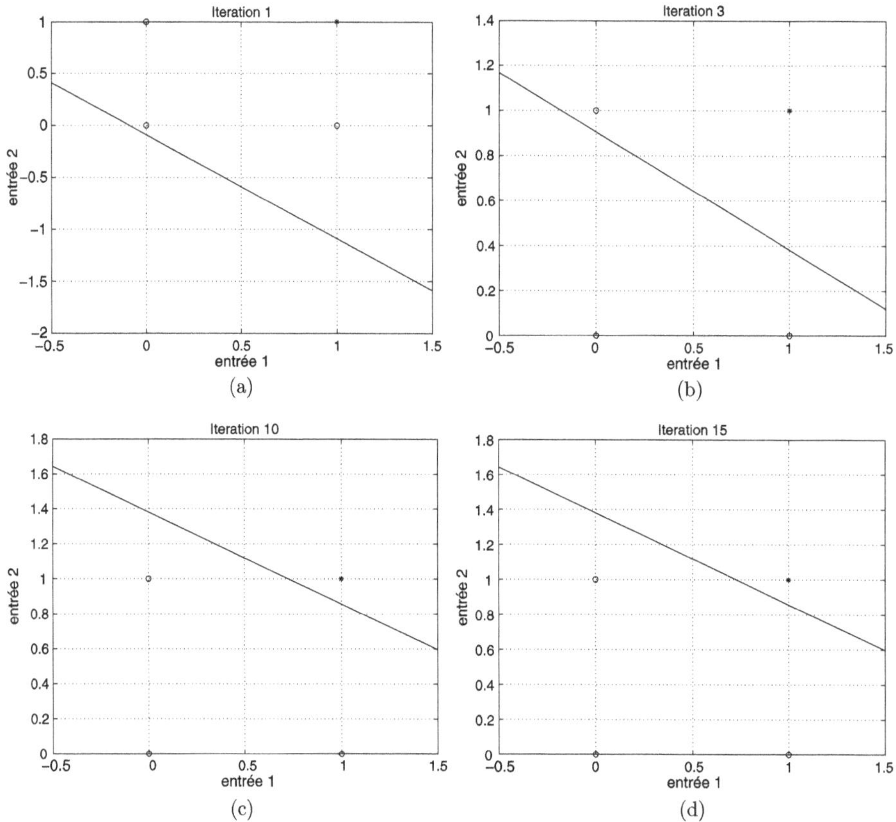

Figure 1.9: Training and separability for the modelling of the AND gate.

obtained after epoch 6. A single neuron can solve a linearly separable problem, where it will struggle to solve a nonlinear one given, for example, by an inclusive gate of the XOR type. This is imposible as a single neuron can only implements a single line and is, therefore, not able to efficiently separate 1 s from 0 s.

1.8 Linear combiner and the LMS algorithm

Following the proven limitation of the Perceptron stated earlier for nonlinearly separable problems, researchers came back to fundamentals. In this section, the linear combiner (which is in fact the linear part of a formal neuron), along with the Least Mean Square (LMS) training algorithm is presented as a prelude for subsequent neural learning rules [11].

A linear adaptive combiner is given in Figure 1.10. As it can be seen, it is constituted of the linear part of a formal neuron (e. g., Perceptron), where inputs are affected by weights w_i, without being affected by a nonlinear transformation at the output (no activation function).

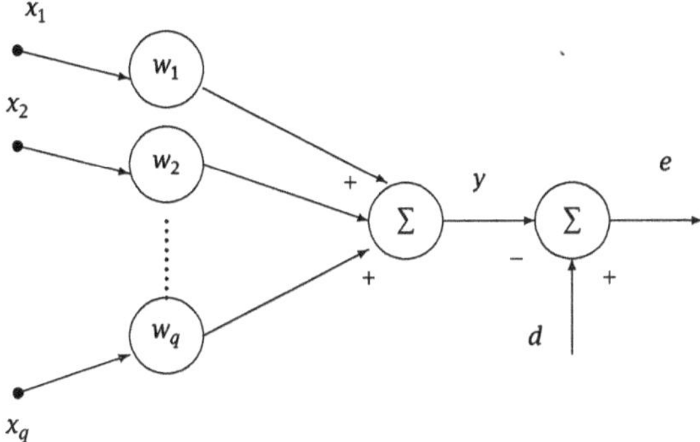

Figure 1.10: Linear combiner.

The output error in then given by

$$e_k = d_k - X^T W \tag{1.12}$$

when both parts of the equation are squared,

$$e_k^2 = d_k^2 + W^T X_k X_k^T W - 2 d_k X_k^T W \tag{1.13}$$

$$E[e_k^2] = E[d_k^2] + W^T E[X_k X_k] W - 2E[d_k X_k^T] W \tag{1.14}$$

if $R = E[X_k X_k^T]$, and $P = E[d_k X_k^T]$, the MSE is given by:

$$\text{MQE} = \xi = E[e_k^2] = E[d_k^2] + W^T R W - 2P^T W \tag{1.15}$$

Remarks:
- It is assumed that X_k and d_k are stationary signals.
- ξ is a quadratic function of the weights and is defined by the performance surface.
- The optimal surface is a hyper-parabolic.
- W^* is the ideal weight associated to the best Mean Squared Error (MSE).

In these conditions, the optimal weight vector W^* is calculating minimizing ξ deriving for W.

$$\nabla = \frac{\delta\xi}{\delta W} = \left[\frac{\delta\xi}{\delta W_0}\frac{\delta\xi}{\delta W_1}\cdots\frac{\delta\xi}{\delta W_L}\right]^T \tag{1.16}$$

$$\nabla = \frac{\delta\xi}{\delta W} = 2RW - 2P \tag{1.17}$$

Note that

$$\frac{\delta}{\delta x}\left(x^T Ax\right) = 2Ax, \quad \text{and} \quad \frac{\delta}{\delta x}(Bx) = B^T \tag{1.18}$$

derived for W^*, gives

$$\nabla = 0 = 2RW^* - 2P \tag{1.19}$$

and we obtain

$$W^* = R^{-1}P \tag{1.20}$$

This analytic solution is known as the equation of Wiener–Hopf. The minimum Mean Squared Error (MSE) is then given by

$$\xi_{\min} = E\left[d_k^2\right] + W^{*T}RW^* - 2P^TW^* \tag{1.21}$$

$$\xi_{\min} = E\left[d_k^2\right] + P^TR^{-1}P \tag{1.22}$$

Remarks:
- The analytical solution is rarely achievable, as car x and d are rarely known a-priori.
- $\xi_{\min}$ can equal 0, but this is not a necessary condition.
- The mean quadratic error can be given by

$$\xi = \xi_{\min} + V^TRV \tag{1.23}$$

where V is the weight vector transform, which has W^* as origin:

$$V = W - W^* \tag{1.24}$$

1.9 Gradient descent optimisation

This section illustrates the gradient descent optimisation principle that will be used for subsequent neural learning rule (The Delta learning rule). Let us consider the scalar case where only a single weight has to be found. The gradient descent research problem is graphically represented in Figure 1.11.

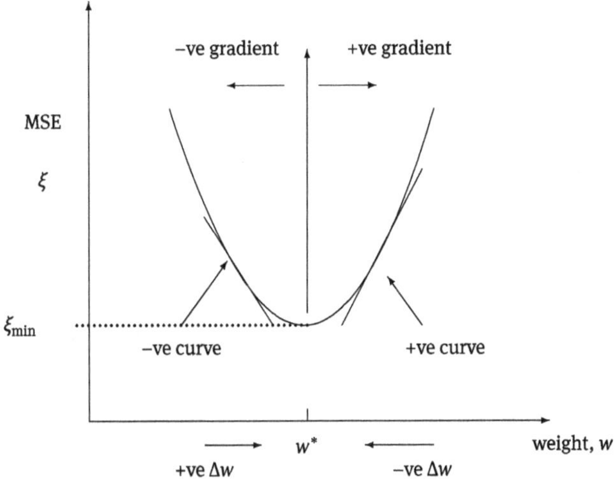

Figure 1.11: Gradient research: the scalar case.

The scalar surface is represented by

$$\xi = \xi_{\min} + \lambda(w - w^*)^2 \tag{1.25}$$

If the following weight update rule is proposed:

$$w_{k+1} = w_k + \mu(-\nabla_k) \tag{1.26}$$

the gradient is obtained by

$$\nabla_k = \left.\frac{d\xi}{dw}\right|_{w=w_k} = 2\lambda(w_k - w^*) \tag{1.27}$$

The update equation is then

$$w_{k+1} = (1 - 2\mu\lambda)w_k + 2\mu\lambda w^* \tag{1.28}$$

This has as a solution

$$w_k = w^* + (1 - 2\mu\lambda)^k(w_0 - w^*) \tag{1.29}$$

$$w_k = w^* + r^k(w_0 - w^*) \tag{1.30}$$

Noting that $|r| < 1$, $r^k \to 0$, when $k \to \infty \Rightarrow w \to w^*$

1.9.1 Stability and learning rate

From equation (1.30), and $w_k \to w^*$, when $k \to \infty$, knowing that

$$|r| = |1 - 2\lambda| < 1 \tag{1.31}$$

in order to guarantee stability:

$$\frac{1}{\lambda} > \mu > 0 \tag{1.32}$$

Remarks:
- Important values of μ will lead to a less stable convergence.
- Smaller values of μ will lead to slower but more stable convergence.

1.9.2 Generalisation to a multi-dimensional system

For multi-dimensional systems, the theory still holds, and the performance system becomes

$$\xi = \xi_{\min} + V^T RV, \quad V = W - W^* \tag{1.33}$$

The update rule is then

$$W_{k+1} = W_k + \mu(-\nabla) \tag{1.34}$$

Note that ∇ is the direction of the steepest descent. The stability condition given for the scalar case, equation (1.32) becomes

$$\frac{1}{\lambda_{\max}} > \mu > 0 \tag{1.35}$$

where $\lambda_{\max}$ is the greatest eigenvalue of the correlation input matrix R.

1.10 LMS learning algorithm

In the general case, the weights update equation for a multi-dimensional system equation (1.34), is rarely achievable as the true gradient depends on W^*, an estimation of the latter is then necessary.

Solution 1. Using an average of the MSE (Mean Squared Error) over a mean horizon, and take the finite difference of the latter.

The principal drawback of this approach is that we need to wait until a given number of error samples to be collected before being used to update the weights.

Solution 2. Instead of using the complete MSE, we restrain ourselves to the Squared Error (SE):

$$\hat{\xi} = e_k^2 \tag{1.36}$$

Now the estimated gradient is given by

$$\hat{\nabla}_k = \begin{bmatrix} \frac{\partial e_k^2}{\partial w_0} \\ \frac{\partial e_k^2}{\partial w_1} \\ \vdots \\ \frac{\partial e_k^2}{\partial w_L} \end{bmatrix} = 2e_k \begin{bmatrix} \frac{\partial e_k}{\partial w_0} \\ \frac{\partial e_k}{\partial w_1} \\ \vdots \\ \frac{\partial e_k}{\partial w_L} \end{bmatrix} = -2e_k X_k \tag{1.37}$$

The update equation becomes then

$$W_{k+1} = W_k + \mu(-\hat{\nabla}) \tag{1.38}$$

$$W_{k+1} = W_k + 2\mu e_k X_k \tag{1.39}$$

Equation (1.39) is the LMS algorithm, and the update convergence of the weight vector, equation (1.39) can be proven [11].

1.10.1 The Adaline

The Adaline (Adaptive Linear Element) is a neuron type of sign type, to which a difference between the linear output z and a desired output d is added, Figure 1.12.

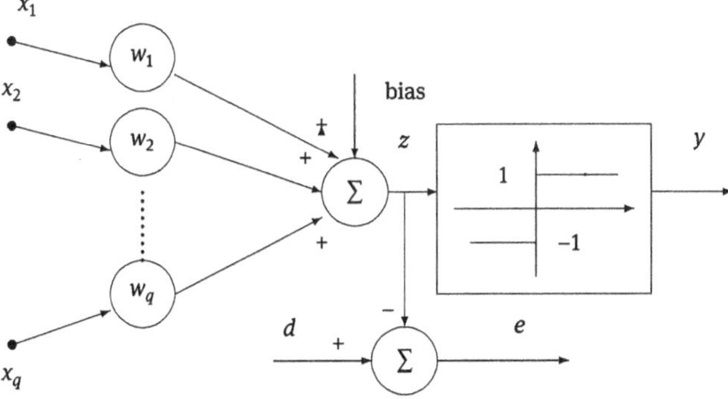

Figure 1.12: Adaline neuron.

Characteristics:
- The linear some z, is used via the LMS algorithm. After training the activation function of type 0/1 is added.

- The weights are then updated using a value proportional to the difference between the actual and desired output.
- The Adaline and the LMS algorithm may give a solution in some cases where, the Perceptron using the perceptron learning rule fails to do so.

1.10.2 The Delta learning rule

The Delta learning rule was introduced in 1986, and uses the LMS algorithm with a Perceptron type neuron [12, 14].

The characteristics of the Delta learning rule are summarised in:
- The Delta rule use only a continuous and differentiable activation function neuron.
- The Delta learning rule is a gradient based method.
- The Delta learning rule is generalised to the Multi-layered Perceptron (MLP).
- The Delta learning rule uses the same approximation of the MSE, used for the LMS algorithm, i. e., the Squared Error (SE).
- The Delta learning rule has no stability measure for the choice of μ.

1.10.3 Derivation of the Delta learning rule

For a continuous Perceptron neuron, we have

$$y = f(z), \quad z = W^T X, \quad e = d - y \tag{1.40}$$

where

$$f_l(z) = \frac{1}{1 + e^{-\beta z}}, \quad f_t(z) = \frac{2}{1 + e^{-\beta z}} - 1 \tag{1.41}$$

with, f_l and f_t corresponding respectively to the log-sigmoidal and tangent-sigmoidal functions.

The gradient of the quadratic error is evaluated as follows:

$$\nabla = \frac{\partial e^2}{\partial W} = -2(d - f(W^T X))f'(W^T X)X \tag{1.42}$$

The weight update is performed as

$$\triangle W = -\mu\nabla \tag{1.43}$$

obtaining the Delta learning rule as

$$W_{k+1} = W_k + 2\mu e f'(z)X \tag{1.44}$$

where $f'(z)$ is the derivative of $f(.)$ evaluated at the operative point, z, as shown in Figure 1.13.

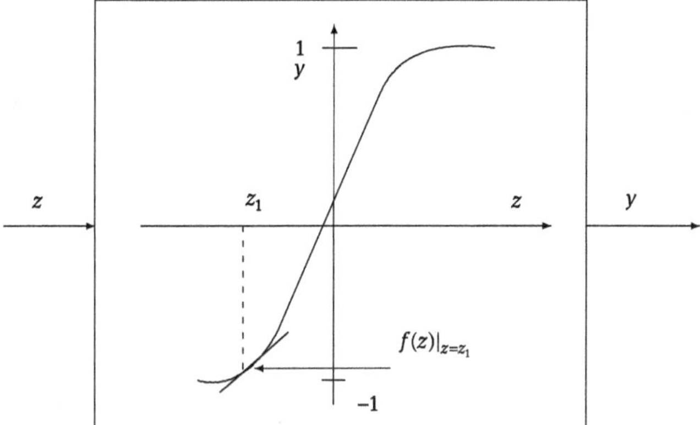

Figure 1.13: Derivation and slope of $f(.)$ during one learning iteration.

1.10.4 Derivative evaluation for the Delta learning rule

If it is assumed a value of $\beta = 1$, during the training procedure without losing generalisation in what follows:

For the unipolar sigmoidal function (log-sigmoidal), we have

$$y = f_l(z) = \frac{1}{1 + e^{-z}} \tag{1.45}$$

$$f_l'(z) = \frac{e^{-z}}{(1 + e^{-z})^2} \tag{1.46}$$

From equation (1.45), we have

$$1 + e^{-z} = \frac{1}{y} \tag{1.47}$$

and

$$e^{-z} = \frac{1}{y} - 1 = \frac{1 - y}{y} \tag{1.48}$$

this gives

$$f_l'(z) = y(1 - y) \tag{1.49}$$

For the bipolar sigmoidal function (tan-sigmoidal), we have

$$y = f_t(z) = \frac{2}{1 + e^{-z}} - 1 \tag{1.50}$$

$$f_t'(z) = \frac{2e^{-z}}{(1 + e^{-z})^2} \tag{1.51}$$

From equation (1.50), we have

$$1 + e^{-z} = \frac{2}{1+y} \tag{1.52}$$

and

$$e^{-z} = \frac{2}{1+y} - 1 = \frac{1-y}{1+y} \tag{1.53}$$

this gives

$$f_t'(z) = \frac{1}{2}(1 - y^2) \tag{1.54}$$

Replacing f_t' in the weights update equation (1.44), we obtain for the unipolar sigmoid function *(log-sigmoid)*:

$$W_{k+1} = W_k + 2\mu e y(1-y)X \tag{1.55}$$

for the bipolar sigmoid function *(tan-sigmoid)*:

$$W_{k+1} = W_k + \mu e(1-y^2)X \tag{1.56}$$

The Delta learning rule is the most sophisticated training algorithm for a single neuron, and constitutes the basis of the backpropagation training rule applied to neural networks, more precisely Multi-layered Perceptrons (MLPs); see Chapter 3.

Bibliography

[1] Hebb, D. O. (1949) The Organization of Behavior, New York, Wiley.
[2] McCulloch, P. and Pitts, W. A. (1943) Logical calculus of ideas immanent in nervous activity, The Bulletin of Mathematical Biophysics, 5, 115–133.
[3] Rosenblatt, F. (1958) The perceptron: a probabilistic model for information storage and organization in the brain. Psychological Review, 65(6).
[4] Rosenblatt, R. (1959) Principles of Neurodynamics. New York: Spartan.
[5] Steinbuch, K. and Widrow, B. (1965) A critical comparison of two kinds of adaptive classification networks, IEEE Transactions on Electronic Computers, 737–740.
[6] Minsky, M. and Papert, S. (1969) Perceptrons: An Introduction to Computational Geometry, MIT Press.
[7] Albus, J. S. (1975) A new approach to manipulator control: the cerebellar model articulation controller (CMAC), in Trans. ASME, Series G. Journal of Dynamic Systems, Measurement and Control, Vol. 97, pp. 220–233.
[8] Hopfield, J. J. (1982) Neural networks and physical systems with emergent collective computational abilities, Proceedings of the National Academy of Sciences, 79(8), 2554–2558.
[9] Meireles, M. R. G., Almeida, P. E. M., and Simoes, M. G. (2003) A comprehensive review for industrial applicability of artificial neural networks. IEEE Transactions on Industrial Electronics, 50(3), 585–601. doi:10.1109/tie.2003.812470.

[10] Abiodun, O. I., Jantan, A., Omolara, A. E., Dada, K. V., Mohamed, N. A., and Arshad, H. (2018) State-of-the-art in artificial neural network applications: a survey. Heliyon, 4(11), e00938. doi:10.1016/j.heliyon.2018.e00938.
[11] Wang, Z.-Q., Manry, M. T., and Schiano, J. L. (2000) LMS learning algorithms: misconceptions and new results on convergence. IEEE Transactions on Neural Networks, 11(1), 47–56. doi:10.1109/72.822509.
[12] Widrow, B. and Hoff, M. E. (1960) Adaptive switching circuits, in 1960 IRE WESCON Conv. Rec., pp. 96–104.
[13] Buhmann, M. D. (2000) Radial basis functions. Acta Numerica, 9(0), 1–38. doi:10.1017/s0962492900000015.
[14] Widrow, B. and Lehr, M. A. (1990) 30 years of adaptive neural networks: perceptron, madaline, and backpropagation. Proceedings of the IEEE, 78(9), 1415–1442.
[15] Ackley, D. H., Hinton, G. E., and Sejnowski, T. J. (1985) A learning algorithm for Boltzmann machines. Cognitive Science, 9, 147–169.
[16] Kosko, B. (1988) Bidirectional associative memories. IEEE Transactions on Systems, Man and Cybernetics, 18(1).
[17] Rumelhart, D. E., Hinton, McClelland, and Williams, R. J. (1986) Learning Internal Representations by Error Propagation Parallel Distributed Processing: Explorations in the Microstructure of Cognition.
[18] Specht, D. F. (1990) Probabilistic neural networks. Neural Networks, 3, 109–118.
[19] Broomhead, D. S. and Lowe, D. (1988) Multivariable functional interpolation and adaptive networks, Complex Systems, 2, 321–355.
[20] Chen, S. and Billings, S. A., (1992) Neural networks for nonlinear dynamic system modelling and identification, International Journal of Control, 56(2), 319–346.
[21] Kohonen, T. (1982) Self-organized formation of topologically correct feature maps, Biological Cybernetics, 46, 59–69.
[22] Yang, T. and Yang, L. B. (1996) The global stability of fuzzy cellular neural network. IEEE Transactions on Circuits and Systems. I, Fundamental Theory and Applications, 43(10), 880–883.

2 Artificial neural networks for food processes: a survey

Advanced and intelligent tools and procedures applied in food processing, could be beneficial in increasing the process efficiency and maintaining the quality requirements of the final product, often drastic and closely related to health issues. Artificial intelligence and its different paradigms, counts among the most popular and fashionable approaches that can, are, and will be used in food processing.

The history of Artificial Intelligence (AI) knew ups and downs during the last 50 years—profitable periods such as the ones from its infancy in the 1950s until the early 1970s, lack of interest and results in the 1980s, and disillusions in the early 1990s as the existing hardware did not match computational needs. With breakthroughs in digitisation of entire economic and social fields, research programmes and innovations picked up, and AI based solutions are nowadays a reality in almost all fields of application.

Figure 2.1 shows the evolution stated above, with the introduction of important AI paradigms at each period. For instance, Expert Systems in the 1980s were designed to simulate specific expert human tasks and were applied to many fields such as medicine, manufacture, etc. In the 1990s, AI enters a perfection era where applications start to beat human performances, with the defeat of the Russian chess champion, Garry Kasparov, by an AI based on a super-calculator baptised "Deep Blue", a breakthrough in human perception and acceptance of AI in 1997 [1].

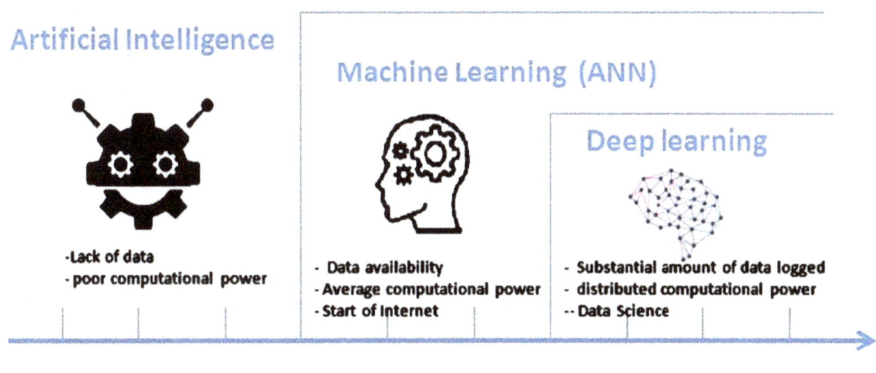

Figure 2.1: Evolution of AI and AI paradigms since 1950.

With a wider usage of the Internet, the world wide web, search engines, and more importantly hardware technologies acquired high performances and quasi-infinite data warehouses. Since the start of the new century, we are witnessing the explosion of

https://doi.org/10.1515/9783110646054-002

new information technologies and revolutionary applications based on infinite data acquisition and new artificial intelligence tools and paradigms such as: image recognition and classification, automatic colouring, automatic subtitles, video indexing, autonomous vehicles, etc.

Artificial Neural Networks (ANNs) are among those tools and paradigms, in this chapter a comprehensive survey of their usage in food processing industry is presented based on most famous classifications.

Due to their ability discussed earlier, especially the fact that they can approximate any nonlinear function (Section 3.5), Artificial Neural Networks (ANNs) have been applied in almost every aspect of food science over the past two and a half decades. However, when compared to other engineering fields, food processes are still behind in terms of number of applications. The main reason is that machine learning and ANNs relies strongly on large amount of data, and its only when the data explosion occurred in modern food processing that solutions required sophisticated analysis methods, in order to: uncover the hidden causal relationships between single or multiple responses and a large set of properties, modelling process behaviour, prediction of processes and food evolution, model based predictive control, etc. One of the most recent articles discussing the use of ANN in the food industry is authored by [22]. The objective of the review was to highlight the application of ANN in the food processing, and evaluate its range of use and adaptability to different food systems referencing more than 50 applications.

Indeed, in food industry, processing operations face problems due to uncertainties and complexities of chemical and biological changes taking place in a particular process due to essentially heating, fermentation, aging, etc.

Huang et al. in 2007 [23], presented a classification of application of ANN in Food Science based on five meta classes cited below, where each class concerns a number of applications in different fields of food processing.

2.1 Machine perception

Machine perception is a promising application fields as it embraces odor, appearance (including shape and colour), and sometimes taste and flavor of a food product. It relies solemnly on sensors gathered data and sensor analysis.

Figure 2.2 shows the steps taken in a machine perception recognition process using ANN as a classification/modeling tool. Based on natural perception, recognition and decision process, the data gathered from food and food based products can be either classified as odors, taste, or vision images given respectively what is known as E-nose, E-tongue, and machine vision perception systems. These three perception approaches of food technology are described in more detail in what follows.

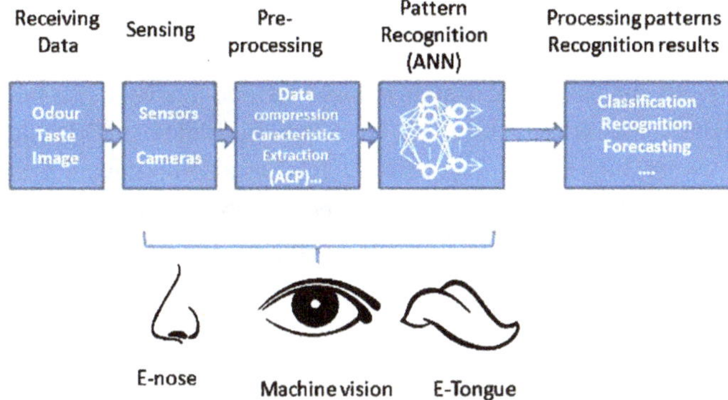

Figure 2.2: Machine perception systems based on ANN: e-nose, e-tongue, and machine vision.

2.1.1 Electronic nose

Electronic nose is an olfactory machine system, providing a relatively recent tool to complement sensory evaluation. The electronic nose is a simulation of human olfactory senses, and has many advantages over biological olfactory capabilities; one of the most obvious is working in dangerous and hazardous environments where humans can be exposed. An electronic nose system consists of a sensor array, a signal conditioning, and data pre-processor, and a pattern recognition system [24].

A comprehensive description of electronic nose is given in [25], and the description of an electronic nose coupled to ANN is given in [26]. Before introducing ANNs as a classifier in the machine perception process, analysis of odour and flavor in food has been traditionally performed either by a trained sensory panel or by headspace gas chromatography mass spectrometry. The authors describe the prerequisite of electronic nose in terms of sensors and data analysis prior to classifier feeding. It presents the most efficient gas sensors in electronic nose instruments. Although at that time ANN usage, especially in food industry, was at its infancy the document well describes the mean of using any type of ANN as a data analysis tool in an electronic nose.

ANNs can then be used as the core of the pattern recognition or modelling system. The electronic nose principle has been developed in various forms and applied in various fields covering classification and quality analysis mainly for various products such as fruits and vegetables, meat and fish, dairy products, oil, grains, beverages, etc. and basically for products with a relatively strong smell where Huang et al., referenced more than 38 applications using this approach. More recently, [27] in 2014 reported more than 60 applications of e-nose using ANNs, all reported in Tables 2.1 to 2.8.

2.1.2 Electronic tongue

Similar to the electronic nose, the electronic tongue focuses on food flavors and taste. The human tongue contains sensors, in the form of 10,000 taste buds of 50 to 100 taste cells each [28], for sour, sweet, bitter, salty, and umami. It relies also on flavour and taste sensors collecting information for the five tastes cited, followed by a treatment and analysis phase that can be based on ANNs. Electronic tongues aim to discriminate and analyse solid foods and beverages based as said on sensing technologies contributing greatly to quality management. Tahara and Toko [29] dresses a comprehensive review of electronic tongue developments and sensor issues. On the other hand, [39] describes more than 40 applications using ANNs as a classifier in an e-tongue approach. More than 10 applications using e-nose for food quality enhancement and assessments for various products such as beer, wine, tea, milk, cooking oil, fruits and vegetables, and meat and fish. Jamal et al., in 2009 [40] focused on E-tongue applications using ANNs precisely, describing more than 12 applications.

More recently, recent advances in Electronic Tongue are presented in [41]. The authors presents a summary of bioelectronic tongues including biomolecule types, transducer part, biological recognition component, and target molecules, where only two applications using ANNs are presented. Sliwinska et al., in 2014 [27], focused on a broader aspect of Food Analysis Using Artificial Senses (mostly olfactive and taste) and reported three applications of E-tongue using ANNs. Winquist, et al., in 1998, monitored milk freshness using an E-tongue approach [61]. Independently, to referenced applications in the aforementioned reviews, the author found more than 25 reported applications of E-tongue using ANNs that will be summarised in subsequent sections.

2.1.3 Machine vision

Machine vision systems in food processing application, is similar to machine vision for other fields such as face and fingerprint recognition in biometrics such as one of the most famous face recognition system “google face”.

In the food science area, machine vision systems accomplish tasks related to visual quality control in order to replace human inspectors in hazardous and fine environments, reduce inspection errors and increase recognition, classification and selection efficiency. A machine vision system may be divided into three distinct parts: an image acquisition part, usually constituted of a camera, an image processing part, and a pattern recognition part; see Figure 2.2. The pattern recognition system may be based on an ANN and more recently deep ANNs most precisely Convolutional Neural Networks (CNNs). The most recent survey on the matter, investigating Deep learning in agriculture is found in [42], where not less than 40 research references that employ deep learning techniques, applied to various agricultural and food production challenges are described and classified.

Classification, clustering, recognition and forecasting a given food image, requires extracting image features first using many mathematical texture and data reduction tools such as PCA (Principal Component Analysis). In the case of Convolutional Neural Networks (CNNs), the extraction step is performed within the first few layers of the network.

As for E-nose and E-tongue, the process is termed pattern recognition and ANNs are one of the most popular tools for the task concerning machine vision food applications. A textbook example of the way how machine vision is applied to food analysis is provided in the work of Ding and Gunasekaran (1994) [43] where a textbook selection system of damaged and undamaged corn kernels, almonds, and animal crackers based on a machine vision is described in the steps below:

- First, 144 digital images of damaged and undamaged corn kernels, almonds, and animal crackers are acquired.
- Image features are then extracted. An average reference shape for undamaged objects is constructed.
- The shape of the inspected object is then compared with a given reference shape. The resulting evaluated or calculated difference, using some similarity and distance approaches, is used as an input vector for an ANN classifier.

The results showed that with an ANN for pattern recognition up to 98 % classification accuracy was achieved.

Similar to those results, the application of ANNs in the machine vision food industry proved its efficiency where 34 applications are reported in [23] among them 11 uses ANNs. In more specific reviews, on ANNs machine vision applications in food industry Cubero et al. in 2011 [44] presented a specific review for quality evaluation of fruits and vegetables. More than 40 reported applications are presented and discussed with 6 applications using ANNs as a classifier. For instance, even a 3D approach is used in [45] for oranges classification.

Davies in 2009 [48] presents a broader review for machine vision applied to food and agriculture. More than 60 applications were reported and many reviews and textbook on state of the art machine vision systems in food industry referenced, where only two (2) applications used ANNs. An earlier review from Du and Sun (2006) [49], focusses mainly on statistical learning applications for classification of machine vision food products with more than 30 applications reported. Other machine learning techniques for food quality evaluation using computer vision (Fuzzy logic, decision tree, and genetic algorithms) are presented with more than 20 applications. Finally, ANNs are discussed as a perspective and only 10 reported applications are presented.

A substancious and representative example using ANN is given by [50] for the prediction of snack sensory quality through image features. The approach used 100 samples of snack images obtained from 600 snack bags, with a charged couple device colour camera. Not less than nine morphology and thirteen texture features were obtained from the images, to be used as an input vector for the neural network. For the output data and reference targets, seven sensory attributes defined by a sensory panel were used.

The obtained results accuracy exceeded 91 %, and remains strongly related to the sensory attributes and the raw material conditions of the snack bags. Note that in order to decrease the model size, the twenty-two input vector dimension was compressed to eight or eleven features. However, decreasing the input dimensions affected the prediction accuracy, which decreased to above 50 %, and 46 % for eight and eleven input features, respectively.

Another ANN reported application, is the usage of a multi-linear regression model to classify poultry carcass, which achieved a prediction accuracy of 82 to 96 % for separation of normal to abnormal classes or classification of two abnormal classes. Surprisingly, the linear model outperformed slightly the ANN ones, proving the heuristic nature of the latter.

Ghazanfari et al. in 1997 [51] graded pistachio nuts into four classes (each class consisting in 260 samples), through the use of images acquired with a frame grabber and RGB video camera.

Deep learning architectures used in a machine vision system proved recently their efficiency due to their ability to well handle multi-modal data. Coming back to [42], one counts 7 applications on raw food while the rest concerns plants and farming issues. For instance, Amara et al. in 2017 [52], proposes a Deep Learning (DL) based approach for Banana Leaf Diseases Classification with very good recognition accuracy rate. Rahnemoonfar and Sheppard, (2017) [53], Zhang et al., (2017) [55], and Sa et al., (2016) [54] investigated fruit counting and detection, respectively, based on deep simulated learning neural networks. For cooked meals and food classification as well as recognition using CNNs, many references can be cited. The most recent and popular references are given in [57, 58].

Undoubtedly, E-nose, E-tongue, and machine vision provide new approaches to a sensory evaluation, automated quality control and remote sensing, many improvements are still to be made in the development and implementation of these systems when it comes to the food industry. However, the development of highly sensitive sensor array and image acquisition systems easing effective feature extraction, and simplifying pattern recognition systems, coupled with greater computing power, the future looks brighter, and the combination of electronic nose, electronic tongues, and machine vision should become more popular and realisable in practice.

2.2 Spectroscopic data interpretation

Quantitative analysis and identification of functional groups based on spectral data interpretation may also be performed using ANNs. Huang et al., [23] cited 7 references in that field concluding that the feedforward MLP trained using backpropagation is the most commonly used neural network for spectral data interpretation. The input data fed to the ANN classifier are spectral data interpretation, measurement of absorption intensity, or compressed spectral information. Meanwhile, the analyte concentration, or a desired physical or chemical characteristic constitutes the output vector.

Not less than 13 applications are discussed in [23] where ANNs are used for quantitative analysis using spectral data. The applications vary from the prediction of sugar content in Golden apples [59], the classification of olive oil [30], the detection of adulteration in freeze-dried instant coffee by Fourier transform infrared spectroscopy with feedforward neural network models [31], the identification of wheat variety with mass spectrometry and an ANN [33], the detection of poultry septicemic livers involving visible/NIR spectroscopy [34].

Apart from the discussed application, [23] concludes on some important issues of using ANNs using spectral data, such as data compression to avoid using large neural networks. Such compressions might be obtained using PCA (Principal Component Analysis). Applying PCA for data compression generally can improve the ANN performance with the risk of discarding some important features if the compression is too high. If the user target is to use all the available data and feature, he is then better off using deep architectures such as CNNs or Stacked Denoising Autoencoders; see Chapter 6.

2.3 Food microbiology and food fermentation

The search of a global universal model in microbial growth for the replacement of the expensive and time-consuming microbial numerical techniques was always a goal in order to predict microbial growth in various food and food product, a key factor for shelf-life and marketing strategies.

Linear models, such as Box and Jenkins [32], are a traditionally used technique to model and predict microbial growth, approximating the nonlinear nature of the phenomena by a linear approach. This has obvious advantages, such as a well-mastered algebraic approach; however, inducing important errors when the microbial growth is highly nonlinear.

Most recently, ANNs were used as an alternative method for modelling microbial growth. As in other fields, a majority of publications claim that ANN models achieve better agreement with experimental data than linear ones, [23] cite six references comforting this thesis.

The developed ANN models can be developed to predict either:

- directly the number of microbes or the microbe growth rate, as in [35];
- predicting, indirectly, the parameters of an existing model, as in [36].

Huang et al., 2007 [23] depict and describe 10 applications of ANNs in food microbiology and fermentation, such as: the growth of Listeria monocytogenes grown in tryptic meat broth [35], prediction of the age of sherry wine vinegars [37], and classification of microbial defects in milk inoculated with different microbial species and stored for different time periods, [38]. For fresh-cut fruits and vegetables, a recent survey of Ma et al., in 2017 describes the most recent developements in novel shelf life extention [56].

2.4 Other applications

Apart from the above classification, neural networks are being, and can be used in almost every aspect of food science and, therefore, be classified outside the aforementioned classes. In food analysis, predicting the functionality, neural networks can translate specific pattern, rheological, physical, chemical properties of various food products, etc.

In the next section, a broader classification and summary of ANN applications classified by food nature may give the reader a shortcut to their field of applications as well as some perspectives.

2.5 Overview of ANN applications and perspectives classified by food types

As presented in earlier sections, artificial neural networks have applications in every aspect of food science and is drawing increasing attention from the community as technology improves and the field matures when more data and process information are collected.

Currently, most applications of ANNs in food science permitted to solve or help solving and contributed to better aided design processes, and reached limits in machine learning design. The next step is undoubtedly the experimentation of deep learning and deep neural networks.

In the following subsections, a comprehensive summary of ANNs applications in food science is given for different food categories.

2.5.1 Dairy and dairy products

The usefulness of ANN models is proven in many dairy applications such as: prediction of milk shelf life, flavor-related shelf life, temperature control, the changes in the physical, chemical, and microbiological structure of yogurt, quantitative determination of protein, seasonal variations in the fatty acids composition of butters, selection of type of cheeses, etc.

Table 2.1: ANN applications in dairy products.

Dairy products	Description	Approach used	ANN type	References
MILK	Shelf life milk estimation		MLP	[2, 3]
	Shelf life of milk based sweetmeat		RBF	[4]
	Solubility index of roller dried goat whole milk		RBF, MLP	[5–7]
	Modelling of a PHE pasteuriser		MLP	[8]
	Milk production estimates		MLP	[12]
	PCDD/PCDF concentrations found in human milk and food samples		Kohonen Map	[13]
	monitoring freshness of milk stored at room temperature	E-tongue	MLP	[60, 62]
	discrimination of different types of fermented milk	E-tongue	MLP	[60]
YOGURT	prediction of shelf-life of low-fat yogurt		MLP	[9]
	Authenticity of low-fat yogurt		MLP	[10]
	Quantitative determination of protein		MLP	[11]
BUTTER	Fatty acids were analysis		MLP	[14]
	butter manufacture		MLP	[15, 16]
CHEESE	rheological properties of cheeses		MLP	[17, 18]
	shelf-life of processed cheese		MLP, RBF	[20, 21, 19]

Table 2.1 presents a comprehensive summary of ANN applications in dairy products. Some are already referenced in earlier sections as E-nose and E-tongue applications but most of them are given here.

It is clear that the most treated product is raw milk, mainly estimation of shelf-life, production, monitoring freshness, temperature modelling, etc. This has obvious economic reasons, as milk production has the smallest benefit margins and is of paramount importance as a nutritive value.

2.5.2 Meat and fish

Although the number of applications in the meat and fish industry is restricted, pork and fish retains the scientist's interest due to small benefit margins for pork industry and freshness/short shelf-life for fish and seafood, Table 2.2. The usage of ANNs in the field have a lot of development margin and many aspects may be investigated in the future.

Table 2.2: ANN applications in fish and meat.

Meat and Fish	Description	Approach used	ANN type	References
PORK	Iberian ham determination of the degree of spoilage in ham	E-nose	PNN	[63]
	discrimination of different types of ham	E-nose	PNN,	[69]
	quality analysis of pork tenderloin based on colour identification	Machine vision	MLP,	[72]
FISH AND SEAFOOD	freshness evaluation in samples: Shrimp and cod roe	E-nose	MLP	[64]
	discrimination of samples based on storage time	E-nose	MLP	[65]
	discrimination of storage time; predicting parameters of spoilage of farmed gilt-head seabream	E-tongue	MLP,	[70, 71]
	evaluation of colour to determine the water content in dehydrated shrimp	Machine vision	MLP,	[75]
SAUSAGES	monitoring of the sausage fermentation process	E-nose	MLP	[66]
POULTRY	identification of tumors and bruised skin of poultry filets	Machine vision	MLP,	[73]
BEEF	colour, shape, and structure analysis of beef steak	Machine vision	MLP,	[74]

2.5.3 Fruits and vegetables

Fruits and vegetables are disparately treated using ANNs. Indeed, Table 2.3 presents the very few existing applications, basically using e-nose and machine vision approaches. One may also notice the start of using deep ANNs for fruits and vegetables counting.

2.5.4 Cereals and grains

Very few applications using ANNs for grains and cereals are reported in the literature, and mainly concerns discrimination and verification. Table 2.4 reports only five found applications, mainly for cereals.

Table 2.3: ANN applications for fruits and vegetables.

Fruits and Vegs	Description	Approach used	ANN type	References
TOMATOS	monitoring of the puree spoilage and quality evaluation of tomatoes	E-nose	SOM	[76, 63]
CARROTS	classification of carrots based on structure	Machine vision	MLP	[86]
APPLES	discrimination of apple varieties and types of apples	E-nose	MLP	[79]
	Real time defects detection in green apples	Machine vision	MLP	[46, 47]
ORANGES	post harvest quality evaluation, apples, peaches, or oranges	E-nose	ANN	[80, 81]
	analysis of shape, structure, and surface roughness of Iyokan oranges	Machine vision	MLP	[85]
PEARS	indicators for predicting quality pears	E-nose	MLP	[82, 83]
	identification of pear and pear peduncle shapes	Machine vision	MLP	[84]
OTHERS	Fruit counting based on deep simulated learning	Machine vision	Deep ANN, CNN	[53]
	discrimination of varieties ripeness level, shelf-life, and storage conditions	E-nose	MLP	[77, 78]

Table 2.4: ANN applications for cereals and grains.

Cereal and grains	Description	Approach	ANN type	References
OATS RYE BARLEY	discrimination of samples in relation to the presence of fungi and bacteria in oats, rye, barley	E-nose	MLP	[87]
GRAINS	discrimination of different samples and smell based classification of grains	E-nose	MLP	[88, 89]
RICE	verification of spoiled grains in rice	Machine vision	MLP	[90]
CEREALS	Identify 14 crop species and 26 diseases		Deep ANN, GoogleNet CNN	[149]

2.5.5 Tea

Table 2.5 reports nine applications mainly for classification using all types of ANNs, for instance RBF, SOM, and MLP. One can notice also the usage of TDNN (Time delay

Table 2.5: ANN applications for tea.

Tea	Description	Approach	ANN type	References
FERMENTATION	determination of the optimal duration of fermentation	E-nose	TDNN, SOM	[91]
CLASSIFICATION	discrimination of the quality classes in Longjing tea	E-nose	ANN	[92, 93]
	discrimination of the green tea brands	E-nose	MLP	[94]
	classification of teas characterised by varying quality, regions, and brands	E-nose	MLP, RBF, SOM	[95–97]
	quality discrimination of green tea	E-tongue	MLP	[98]
DISCRIMINATION BLACK TEA	discrimination of black teas based on tea type and brand	E-tongue	MLP	[99]
ANALYSIS	determination of the a flavin level	E-tongue	MLP	[100]

ANNs) for fermentations and it is a process strongly related to time, in a very nonlinear manner.

2.5.6 Alcoholic beverages

The usage of ANNs for alcoholic beverage concerns mainly wine production, detection of origin, identification, and monitoring of aroma change. Table 2.6 reports the few existing applications.

Table 2.6: ANN applications for alcoholic beverages.

Alcohols	Description	Approach	ANN type	References
WINES	detection of falsification of Italian wine	E-nose	MLP	[101]
	botanical and geographical origin, aging process, and method	E-nose	MLP, PNN	[102, 103]
	monitoring of changes in red wine aroma after bottle opening	E-nose	SOM	[65]
	discrimination of wines based on geographical origin	E-tongue	MLP	[104]
	discrimination of wines based on geographical origin; discrimination of wines based on vintage, vineyard, and age	E-tongue	ANN	[105–107]
BEER	monitoring of changes during the aging process in beer	E-nose	MLP, RBF	[108]

2.5.7 Oil

Oil and more precisely olive oil, is much more treated using ANNs, as can be seen in Table 2.7. More than 90 % of the applications concerns olive oil and treat mainly classification, identification, authentication, and traceability issues. The main reason for focusing on olive oil is the strong growing competition between different producing countries and origins. An application for olive oil origin classification is given later in the document: Section 4.4.

Table 2.7: ANN applications for table oil.

Oil	Description	Approach used	ANN type	References
OLIVE OIL	detection of falsification of olive oil	E-nose	MLP	[109]
	discrimination of geographical varieties	E-nose	CP-ANN	[110]
	discrimination of quality classes based on qualitative and quantitative information	E-nose	SOM	[112]
	discrimination of extra-virgin olive oil based on geographical origin and bitterness level	E-tongue	MLP	[113]
	Determination of the geographical origin of Italian extra virgin olive oil using pyrolysis mass spectrometry and artificial neural networks		MLP	[132]
	Classification of olive oils from nine different olive growing regions in Italy		FFNN	[133]
	Authentication of virgin olive oils of very close geographical origins based on NIR spectroscopy		MLP	[134]
	High throughput flow 1H NMR fingerprinting, for classification of geographic origin and year of production		PNN	[135]
	A classification tool for geographical traceability of virgin olive oils		MLP	[136]
	Identification of geographical origin of Spain, Italy and Portugal virgin olive oil using complete chemical characterisation of samples		MLP	[137]
	Traceability of olive oil of Italy, Spain, France, Greece, Cyprus, and Turkey based on volatiles pattern using ANN		MLP	[138]

Table 2.7: (continued)

Oil	Description	Approach used	ANN type	References
	Identification of geographical origin based on High resolution NMR characterisation of olive oils		PNN	[139]
	Traceability of extra virgin olive oils of Spain, Italy, Greece, and Argentina using laser-induced		MLP	[140]
	Breakdown Spectroscopy (LIBS) and neural networks			
	Identification of geographical origin: distinguishing similar EVOO samples from four different but close origins in Spain		MLP	[141]
	Detection of adulterations of extra virgin olive oil with inferior edible oils		SOM	[142]
	Sensory evaluation of virgin olive oils based on the volatile composition data		BP-ANN	[147]
	Electronic nose based on metal oxide semiconductor sensors as a fast alternative for the detection of adulteration of virgin olive oils	E-nose	BP-ANN	[144]
	An electronic nose for differentiation of Extra virgin olive (EVO) and non-virgin olive oil (OI)	E-nose	RBF	[145]
	Virgin Olive Oil Quality Classification Combining Neural Network and MOS Sensors		MLP	[146]
	A sensor-software based on artificial neural network for the optimisation of olive oil elaboration process		FFBP-ANN	[148]
CORN OIL	detection of corn oil in adulterated sesame seed oil	E-nose	PNN	[111]
VEGETABLE OIL	Applied ANN to the signals generated by an electronic nose for the classification of vegetable oils	E-nose	BP-ANN	[143]
	Quantification of Phenolic Compounds in olive oil mill wastewater by artificial neural network		MLP	[141]

2.5.8 Other products

Table 2.8 presents some unclassified products, where coffee and beverages are the most treated, using mainly MLPs. One may notice the usage of RBFs and even deep ANNs for food classifications in restaurants as classification is based on meal images.

Table 2.8: ANN applications for unclassified products.

Other products	Description	Approach used	ANN type	References
VINEGAR	identification of some commercial vinegars	E-nose	MLP	[114]
SPICES	discrimination of different spices	E-nose	MLP	[115]
COFFEE	discrimination of coffee brands, different quality criteria, and bean ripening time	E-nose	ANN	[116–120]
HONEY	discrimination based on botanical origin	E-tongue	MLP	[121, 122]
	discrimination based on botanical and geographical origin	E-tongue	MLP	[123, 124]
BEVERAGE	discrimination between beverages of the same type and monitoring of aging in juice	E-tongue	MLP, SOM	[125–127]
	discrimination between high mineralised and low mineralised waters	E-tongue	SOM	[128]
	discrimination of juice brands: orange, pear, peach, apricot juices	E-tongue	ANN	[129–131]
EGG	determination of egg freshness based on storage time at room temperature	E-nose	BPNN, SOM,	[67, 68]
OTHER	description of an electronic nose coupled to ANN	E-nose	MLP	[26]
	Modelling Restaurant Context for Food Recognition	Machine vision	Deep ANN, CNN	[150]

Bibliography

[1] Chellapilla, K. and Fogel, D. B. (1999) Evolving neural networks to play checkers without relying on expert knowledge, IEEE Transactions on Neural Networks, 10(6).

[2] Vallejo-Cordoba, B., Arteaga, G. E., and Nakai, S. (1995) Predicting milk shelf-life based on artificial neural networks and headspace gas chromatographic data. Journal of Food Science, 60, 885–888.

[3] Doganis, P., Alexandridis, A., Patrinos, P., and Sarimveis, H. (2006) Time series sales forecasting for short shelf-life food products based on artificial neural network models and evolutionary computing. Journal of Food Engineering, 75, 196–204.

[4] Boishebert, V. D., Urruty, L., Giraudel, J. L., and Montury, M. (2004) Assessment of strawberry aroma through solid-phase microextraction-gas chromatography and artificial neuron network methods. Variety classification versus growing years. Journal of Agricultural and Food Chemistry, 52(9), 2472–2478.

[5] Goyal, S., Sitanshu, K., and Goyal, G. K. (2013). Artificial neural networks for analyzing solubility index of roller dried goat whole milk powder. International Journal of Mechanical Engineering and Computer Applications, 1(1), 1–4.

[6] Goyal, S. and Goyal, G. K. (2012) Predicting solubility index of roller dried goat whole milk powder using Bayesian regularization ANN models. Scientific Journal of Zoology, 1(3), 61–68.

[7] Goyal, S. and Goyal, G. K. (2012) Multilayer ANN models for determining solubility index of roller dried goat whole milk powder. Journal of Bioinformatics and Intelligent Control, 1, 76-79.

[8] Khadir, M. T. and Ringwood, J. V. (2001) Neural network modelling and predictive control of milk pasteurisation, in Proc. of the International Conference on Engineering Applications of Neural Networks, Cagliari, Italy, pp. 116–122.

[9] Sofu, A. and Ekinci, F. Y. (2007) Estimation of storage time of yogurt with artificial neural network modeling. Journal of Dairy Science, 90(7), 3118–3125.

[10] Cruz, A. G., Walter, E. H. M., Cadena, R. S., Faria, J. A. F., Bolini, H. M. A., and Fileti, A. M. F. (2009) Monitoring the authenticity oflow-fat yogurts by an artificial neural network. Journal of Dairy Science, 92(10), 4797–4804.

[11] Khanmohammadi, M., Garmarudi, A. B., Ghasemi, K., Garrigues, S., and Guardia, M. (2009) Artificial neural network for quantitative determination of total protein in yogurt by infrared spectrometry. Microchemical Journal, 91(1), 47–52.

[12] Sanzogni, L. and Kerr, D. (2001) Milk production estimates using feed forward artificial neural networks. Computers and Electronics in Agriculture, 32, 21–30.

[13] Nadal M., Espinosa G., Schuhmacher M., and Domingo J. L. (2004) Patterns of PCDDs and PCDFs in human milk and food and their characterization by artificial neural networks. Chemosphere 54, 1375–1382.

[14] Gori, A., Chiara, C., Selenia, M., Nocetti, M., Fabbri, A., Caboni, M. F., and Losi, G. (2011) Prediction of seasonal variation of butters by computing the fatty acids composition with artificial neural networks. European Journal of Lipid Science and Technology, 113(11), 1412–1419.

[15] Funahashia, H. and Horiuchib, J. (2008) Characteristics of the churning process in continuous butter manufacture and modelling using an artificial neural network. International Dairy Journal, 18, 323–328.

[16] Gori, A., Cevoli, C., Fabbri, A., Caboni, M. F., and Losi, G. (2012) A rapid method to discriminate season of production and feeding regimen of butters based on infrared spectroscopy and artificial neural networks. Journal of Food Engineering, 109, 525–530.

[17] Ni, H. and Gunasekaran, S. (1998) Food quality predication with neural networks. Food Technology, 52(10), 60–65.

[18] Jimènez-Mãrquez, S. A., Thibault, J., and Lacroix, C. (2005) Prediction of moisture in cheese of commercial production using neurocomputing models. International Dairy Journal, 15, 1156–1174.

[19] Goyal, S. and Goyal, G. K. (2013) Intelligent artificial neural network computing models for predicting shelf life of processed cheese. Intelligent Decision Technologies, 7(2), 107–111.

[20] Goyal, G. K. and Goyal, S. (2013) Cascade artificial neural network models for predicting shelf life of processed cheese. Journal of Advances in Information Technology, 4(2), 80–83.

[21] Goyal, S. and Goyal, G. K. (2012). Application of simulated neural networks as non-linear modular modeling method for predicting shelf life of processed cheese. Jurnal Intelek, 7(2), 48–54.

[22] Guiné, R. P. F. (2019) The use of artificial neural networks (ANN) in food process engineering, International Journal of Food Engineering 5(1).

[23] Huang, Y., Kangas, L. J., and Rasco, B. A. (2007) Applications of artificial neural networks (ANNs) in food science. Critical Reviews in Food Science and Nutrition, 47(2), 13–26.

[24] Arshak, K., Lyons, G., Cunniffe, C., Harris, J., and Clifford S. (2003) A review of digital data acquisition hardware and software for a portable electronic nose. Sensor Review, 23(4), 332–344.

[25] Estakhroyeh, H. R., Rashedi, E., and Mehran, M. (2018) Design and construction of electronic nose for multi-purpose applications by sensor array arrangement using IBGSA. Journal of Intelligent & Robotic Systems, 92, 205–221.

[26] Haugen, J. E. and Kvaal, K. (1998) Electronic nose and artificial neural network. Meat Science, 49, S273–S286.

[27] Ŝliwiñska, M., Wiŝniewska, P., Dymerski, T., Namieŝnik, J., and Wardencki, W. (2014) Food analysis using artificial senses. Journal of Agricultural and Food Chemistry, 62(7), 1423–1448.

[28] Mohapatra, P. and Panigrahi, S. (2006) Artificial taste sensors: An overview. ASABE/CSBE North Central Intersectional Meeting. Paper Number: MBSK 06-204.

[29] Tahara, Y. and Toko, K. (2013) Electronic tongues–a review. IEEE Sensors Journal, 13(8), 3001–3011.

[30] Angerosa, F., Giacinto, A. D., Vito, R., and Cumitini, S. (1996) Sensory evaluation of virgin olive oils by artificial neural network processing of dynamic headspace gas chromatographic data. Journal of the Science of Food and Agriculture, 72, 323–328.

[31] Briandet, R., Kemsley, E. K., and Wilson, R. H. (1996) Approaches to adulteration detection in instant coffees using infrared spectroscopy and chemometrics. Journal of the Science of Food and Agriculture, 71, 359–366.

[32] Makridakis, S. and Hibon, M. (1997) ARMA models and the Box–Jenkins methodology. Journal of Forecasting, 16(3), 147–163.

[33] Bloch, H. A., Kesmir, C., Petersen, M., Jacobsen, S., and Søndergaard, I. (1999) Identification of wheat varieties using matrix-assisted laser desorption/ionisation time-of-flight mass spectrometry and an artificial neural network. Rapid Communications in Mass Spectrometry, 13, 1535–1539.

[34] Hsieh, C., Chen, Y. R., Dey, B. P., and Chan, D. E. (2001) Separating septicemic and normal chicken livers by visible/near-infrared spectroscopy and backpropagation neural networks. Transactions of the ASABE, 45(2), 459–469.

[35] Cheroutre-Vialette, M., and Lebert, A. (2000) Modelling the growth of Listeria monocytogenes in dynamic conditions. International Journal of Food Microbiology, 55, 201–207.

[36] García-Gimeno, R. M., Hervás-Martínez, C., Sanz-Tapia, E., and Zurera-Cosano, G. (2002) Estimation of microbial growth parameters by means of artificial neural networks. Food Science and Technology International, 8(2), 73–80.

[37] Parrilla, M. C. G., Heredia, F. J., and Troncoso, A. M. (1999) Sherry wine vinegars: phenolic composition changes during aging. Food Research International, 32, 433–440.

[38] Horimoto, Y., Lee, K., and Nakai, S. (1997) Classification of microbial defects in milk using a dynamic headspace gas chromatograph and computer-aided data processing. 2. Artificial neural networks, partial least squares regression analysis, and principal component regression analysis. Journal of Agricultural and Food Chemistry, 45, 743–747.

[39] Baldwin, E. A., Bai, J., Plotto, A., and Dea, S. (2011) Electronic noses and tongues: applications for the food and pharmaceutical industries. Sensors, 11(5), 4744–4766.

[40] Jamal, M., Khan, M. R., and Imam, S. A. (2009) Electronic tongue and their analytical application using artificial neural network approach: a review, MASAUM Journal Of Reviews and Surveys, 1(1).

[41] Ha, D., Sun, Q., Su, K., Wan, H., Li, H., Xu, N., Wang, P. (2015) Recent achievements in electronic tongue and bioelectronic tongue as taste sensors. Sensors and Actuators. B, Chemical, 207, 1136–1146.

[42] Kamilaris, A. and Prenafeta-Boldú, F. X. (2018) Deep learning in agriculture: a survey. Computers and Electronics in Agriculture, 14(7), 70–90.

[43] Ding, K. and Gunasekaran, S. (1994) Shape feature extraction and classification of food material using computer vision. Transactions of the ASAE, 37(5), 1537–1545.

[44] Cubero, S., Aleixos, N., Moltó, E., Gómez-Sanchis, J., and Blasco, J. (2010) Advances in machine vision applications for automatic inspection and quality evaluation of fruits and vegetables. Food and Bioprocess Technology, 4(4), 487–504.

[45] Kanali, C., Murase, H., and Honami, N. (1998) Three-dimensional shape recognition using a charge-simulation method to process primary image features. Journal of Agricultural Engineering Research, 70(2), 195–208.

[46] Bennedsen, B. S., Peterson, D. L., and Tabb, A. (2005) Identifying defects in images of rotating apples. Computers and Electronics in Agriculture, 48, 92–102.

[47] El-Masry, G., Wang, N., Vigneault, C., Qiao, J., and ElSayed, A. (2008) Early detection of apple bruises on different background colors using hyperspectral imaging. LWT – Food Science and Technology, 41, 337–345.

[48] Davies, E. R. (2009) The application of machine vision to food and agriculture: a review. TheImaging Science Journal, 57(4), 197–217.

[49] Du, C. J. and Sun, D.-W. (2006) Learning techniques used in computer vision for food quality evaluation: a review. Journal of Food Engineering, 72(1), 39–55.

[50] Sayeed, M. S., Whittaker, A. D., and Kehtarnavaz, N. D. (1995) Snack quality evaluation method based on image features and neural network prediction. Transactions of the ASAE, 38(4), 1239–1245.

[51] Ghazanfari, A., Irudayaraj, J., Kusalik, A., and Romaniuk, M. (1997) Machine vision grading of pistachio nuts using fourier descriptors. Journal of Agricultural Engineering Research, 68(3), 247–252.

[52] Amara, J., Bouaziz, B., and Algergawy, A. (2017) A Deep Learning-based Approach for Banana Leaf Diseases Classification, Lecture Notes in Informatics (LNI), Gesellschaft für Informatik, 79-88.

[53] Rahnemoonfar, M. and Sheppard, C. (2017) Deep count: fruit counting based on deep simulated learning. Sensors 17(4), 905. Ma, L., Zhang, M., Bhandari, B., and Gao, Z. (2017) Recentdevelopments in novel shelf life extension technologies of fresh-cut fruits and vegetables. Trends in Food Science and Technology, 64, 23-38.

[54] Sa, I., Ge, Z., Dayoub, F., Upcroft, B., Perez, T., and McCool, C. (2016) Deepfruits: a fruit detection system using deep neural networks. Sensors 16(8), 12–22.

[55] Zhang, Y-D., Dong, Z., Chen, X., Jia, W., Du, S., Muhammad, K., and Wang, S-H. (2017) Image based fruit category classification by 13-layer deep convolutional neural network and data augmentation, Multimedia Tools and Applications, 78, 3613–3632.

[56] Ma, L., Zhang, M., Bhandari, B., and Gao, Z. (2017) Recent developments in novel shelf life extension technologies of fresh-cut fruits and vegetables. Trends in Food Science & Technology, 64, 23–38.

[57] Kawano, Y. and Yanai, K. (2014) Automatic expansion of a food image dataset leveraging existing categories with domain adaptation, in Proc. ECCV Workshop Transferring Adapting Source Knowl. Comput. Vision, pp. 3–17.
[58] Hassannejad, H., Matrella, G., Ciampolini, P., De Munari, I., Mordonini, M., and Cagnoni, S. (2016) Food image recognition using very deep convolutional networks, in Proceedings of the 2nd International Workshop onMultimedia Assisted Dietary Management – MADiMa 2016.
[59] Bochereau, L., Bourgine, P., and Palagos, B. (1992) A method for prediction by combining data analysis and neural networks: application to prediction of apple quality using near infra-red spectra. Journal of Agricultural Engineering Research, 51, 207–216.
[60] Winquist, F., Holmin, S., Krantz-Rückler, C., Wide, P., and Lündström, I. (2000) A hybrid electronic tongue. Analytica Chimica Acta, 406, 147–157.
[61] Winquist, F., Krantz-Rückler, C., Wide, P., Lündström, I. (1998) Monitoring of freshness of milk by an electronic tongue on the basis of voltammetry. Measurement Science & Technology, 9, 1937–1946.
[62] Wei, Z., Wang, J., and Zhang, X. (2013) Monitoring of quality and storage time of unsealed pasteurized milk by voltammetric electronic tongue. Electrochimica Acta, 88, 231–239.
[63] García, M., Aleixandre, M., and Horrillo, M. C. (2005) Electronic nose for the identification of spoiled Iberian hams, in Spanish Conference on Electron Devices, Tarragona, Spain, New York: IEEE, pp. 537–540.
[64] Winquist, F., Sundgren, H. and Lundstrom, I. (1998) Electronic noses for food control, in Biosensors for Food Analysis, Scott, A. O., Ed., Cambridge, UK: Royal Society of Chemistry, pp. 170–179.
[65] Di Natale, C., Macagnano, A., Davide, F., D'Amico, A., Paolesse, R., Boschi, T., Faccio, M., and Ferri, G. (1997) An electronic nose forfood analysis. Sensors and Actuators. B, Chemical, 44, 521–526.
[66] Eklov, T., Johansson, G., Winquist, F., and Lündström, I. (1997) Monitoring sausage fermentation using an electronic nose, in Methods to Improve the Selectivity of Gas Sensors, Eklov, T., Ed., Licentiate thesis 565; Sweden: Department of Physics and Measurement Technology, University of Linkoping, pp. 4–50.
[67] Dutta, R., Hines, E. L., Gardner, J. W., Udrea, D. D., and Boilot, P. (2003) Nondestructive egg freshness determination: an electronic nose based approach. Measurement Science & Technology, 14, 190–198.
[68] Liu, P. and Tu, K. (2012) Prediction of TVB-N content in eggs based on electronic nose. Food Control, 23, 177–183.
[69] Garcìa, M., Aleixandre, M., Gutièrrez, J., and Horrillo, M. C. (2006) Electronic nose for ham discrimination. Sensors and Actuators. B, Chemical, 11(4), 418–422.
[70] Gil, L., Barat, J. M., Escriche, I., Garcia-Breijo, E., Martínez-Mañez, R., and Soto, J. (2008) An electronic tongue for fish freshness analysis using a thick-film array of electrodes. Mikrochimica Acta, 163, 121–129.
[71] Gil, L., Barat, J. M., Garcia-Breijo, E., Ibañez, J., Martínez-Màñez, R., Soto, J., Llobet, E., Brezemes, J., Aristoy, J. C., and Toldrà, F. (2008) Fish freshness analysis using metallic potentiometric electrodes. Sensors and Actuators. B, Chemical, 131, 362–370.
[72] Lu, J., Tan, J., Shatadal, P., and Gerrard, D. E. E. (2000) Valuation of pork color by using computer vision. Meat Science, 56, 57–60.
[73] Park, B., Chen, Y. R., Nguyen, M., and Hwang, H. (1996) Characterising multispectral images of tumorous, bruised, skin-torn, and wholesome poultry carcasses. Transactions of the ASAE, 39, 1933–1941.
[74] Li, J., Tan, J., and Martz, F. A. (1997) Predicting beef tenderness from image texture features. ASAE Annual International Meeting Technical Papers, St. Joseph, MI, USA.

[75] Mohebbi, M., Akbarzadeh-T, M.-R., Shahidi, F., Moussavi, M., and Ghoddusi, H. (2009) Computer vision systems (CVS) for moisture content estimation in dehydrated shrimp. Computers and Electronics in Agriculture, 69, 128–134.

[76] Sinesio, F., Di Natale, C., Quaglia, G. B., Bucarelli, F. M., Moneta, E., Macagnano, A., Paolesse, R., and D'Amico, A. (2000) Use of electronic nose and trained sensory panel in the evaluation of tomato quality. Journal of the Science of Food and Agriculture, 80, 63–71.

[77] Brezmes, J., Llobet, E., Vilanova, X., Orts, J., Saiz, G., and Correig, X. (2001) Correlation between electronic nose signals and fruit quality indicators on shelf-life measurements with Pink Lady apples. Sensors and Actuators. B, Chemical, 80, 41–50.

[78] Herrmann, U., Jonischkeit, T., Bargon, J., Hahn, U., Li, Q.-Y., Schalley, C. A., Vögel, E., and Vögtle, F. (2002) Monitoring apple flavor by use of quartz microbalances. Analytical and Bioanalytical Chemistry, 372, 611–614.

[79] Xiaobo, Z. and Jiewen, Z. (2008) Comparative analyses of apple aroma by a tin-oxide gas sensor array device and GC/MS. Food Chemistry, 107, 120–128.

[80] Di Natale, C., Macagnano, A., Martinelli, E., Paolesse, R., Proietti, E., and D'Amico, A. (2001) The evaluation of quality of post-harvest oranges and apples by means of an electronic nose. Sensors and Actuators. B, Chemical, 78, 26–31.

[81] Di Natale, C., Macagnano, A., Martinelli, E., Proietti, E., Paolesse, R., Castellari, L., Campani, S., and D'Amico, A. (2001) Electronic nose based investigation of the sensorial properties of peaches and nectarines. Sensors and Actuators. B, Chemical, 77, 561–566.

[82] Zhang, H., Wang, J., and Ye, S. (2008) Prediction of soluble solids content, firmness and pH of pear by signals of electronic nose sensors. Analytica Chimica Acta, 606, 112–118.

[83] Zhang, H., Wang, J., and Ye, S. (2008) Predictions of acidity, soluble solids and firmness of pear using electronic nose technique. Journal of Food Engineering, 86, 370–378.

[84] Ying, Y., Jing, H., Tao, Y., and Zhang, N. (2003) Detecting stem and shape of pears using fourier transformation and an artificial neural network. Transactions of the ASAE, 46, 157–162.

[85] Kondo, N., Ahmad, U., Monta, M., and Murase, H. (2000) Machine vision based quality evaluation of Iyokan orange fruit using neural networks. Computers and Electronics in Agriculture, 29, 135–147.

[86] Brandon, J. R., Howarth, M. S., Searcy, S. W., and Kehtarnavaz, N. (1990) A Neural Network for Carrot Tip Classification; ASAE: St. Joseph, MI, USA, p13.

[87] Jonsson, A., Winquist, F., Schnürer, J., Sundgren, H., and Lundström, I. (1997) Electronic nose for microbial quality classification of grains. International Journal of Food Microbiology, 35, 187–193.

[88] Börjesson, T. (1996) Detection of off-odours in grains using an electronic nose, in Olfaction and Electronic Nose, 3rd International Symposium; s.n.: France: Toulouse.

[89] Börjesson, T., Eklöv, T., Jonsson, A., Sundgren, H., and Schnürer, J. (1996) Electronic nose for odor classification of grains. Cereal Chemistry, 73, 457–461.

[90] Wan, Y. N., Lin, C. M., and Chiou, J. F. (2000) Adaptive classification method for an automatic grain quality inspection system using machine vision and neural network, in Conference Paper 2000 ASAE Annual International Meeting, Milwaukee, WI, USA, pp. 1–19.

[91] Bhattacharya, N., Tudu, B., Jana, A., Ghosh, D., Bandhopadhyaya, R., and Bhuyan, M. (2008) Preemptive identification of optimum fermentation time for black tea using electronic nose. Sensors and Actuators. B, Chemical, 131, 110–116.

[92] Yu, H., Wang, J., Zhang, H., Yu, Y., and Yao, C. (2008) Identification of green tea grade using different feature of response signal from E-nose sensors. Sensors and Actuators. B, Chemical, 128, 455–461.

[93] Yu, H. and Wang, J. (2007) Discrimination of LongJing green-tea grade by electronic nose. Sensors and Actuators. B, Chemical, 122, 134–140.

[94] Chen, Q., Zhao, J., Chen, Z., Lin, H., and Zhao, D.-A. (2011) Discrimination of green tea quality using the electronic nose technique and the human panel test, comparison of linear and nonlinear classification tools. Sensors and Actuators. B, Chemical, 159, 294–300.

[95] Dutta, R., Kashwan, K. R., Bhuyan, M., Hines, E. L., and Gardner, J. W. (2003) Electronic nose based tea quality standardization. Neural Networks, 16, 847–853.

[96] Tudu, B., Jana, A., Metla, A., Ghosh, D., Bhattacharyya, N., and Bandyopadhyay, R. (2009) Electronic nose for black tea quality evaluation by an incremental RBF network. Sensors and Actuators. B, Chemical, 13(8), 90–95.

[97] Roy, R. B., Tudu, B., Shaw, L., Jana, A., Bhattacharyya, N., and Bandyopadhyay, R. (2012) Instrumental testing of tea by combining theresponses of electronic nose and tongue. Journal of Food Engineering, 110, 356–363.

[98] Chen, Q., Zhao, J., and Vittayapadung, S. (2008) Identification of green tea grade level using electronic tongue and pattern recognition. Food Research International, 41, 500–504.

[99] Palit, M., Tudu, B., Bhattacharyya, N., Dutta, A., Dutta, P. K., Jana, A., Bandyopadhyay, R., and Chatterjee, A. (2010) Comparison of multivariate preprocessing techniques as applied to electronic tongue based pattern classification for black tea. Analytica Chimica Acta, 675, 8–15.

[100] Ghosh, A., Tamuly, P., Bhattacharyya, N., Tudu, B., Gogoi, N., and Bandyopadhyay, R. (2012) Estimation of theaflavin content in black tea using electronic tongue. Journal of Food Engineering, 110, 71–79.

[101] Penza, M. and Cassano, G. (2004) Recognition of adulteration of Italian wines by thin-film multisensor array and artificial neural networks. Analytica Chimica Acta, 509, 159–177.

[102] Di Natale, C., Davide, F. A. M., D'Amico, A., Nelli, P., Groppelli, S., and Sberveglieri, G. (1996) An electronic nose for the recognition of the vineyard of a red wine. Sensors and Actuators. B, Chemical, 33, 83–88.

[103] Lozano, J., Arroyo, T., Santos, J. P., Cabellos, J. M., and Horrillo, M. C. (2008) Electronic nose for wine ageing detection. Sensors and Actuators. B, Chemical, 133, 180–186.

[104] Di Natale, C., Paolesse, R., Macagnano, A., Mantini, A., D'Amico, A., Ubigli, M., Legin, A., Lvova, L., Rudnitskaya, A., and Vlasov, Y. (2000) Application of a combined artificial olfaction and taste system to the quantification of relevant compounds in red wine. Sensors and Actuators. B, Chemical, 69, 342–347.

[105] Kirsanov, D., Mednova, O., Vietoris, V., Kilmartin, P. A., and Legin, A. (2012) Towards reliable estimation of an "electronic tongue" predictive ability from PLS regression models in wine analysis. Talanta, 90, 109–116.

[106] Moreno-Codinachs, L., Kloock, J. P., Schöning, M. J., Baldi, A., Ipatov, A., Bratov, A., and Jiménez-Jorquera, C. (2008) Electronic integrated multisensor tongue applied to grape juice and wine analysis. The Analyst, 133, 1440–1448.

[107] Rudnitskaya, A., Delgadillo, I., Legin, A., Rocha, S. M., Costa, A. M., and Simoes, T. (2007) Prediction of the Port wine age using an electronic tongue. Chemometrics and Intelligent Laboratory Systems, 88, 125–131.

[108] Ghasemi-Varnamkhasti, M., Rodríguez-Méndez, M. L., Mohtasebi, S. S., Apetrei, C., Lozano, J., Ahmadi, H., Razavi, S. H., and de Saja, J. A. (2012) Monitoring the aging of beers using a bioelectronic tongue. Food Control, 25, 216–224.

[109] Cerrato Oliveros, M. C., Pérez Pavón, J. L., García Pinto, C., Fernández Laespada, M. E., Moreno Cordero, B., and Forina, M. (2002) Electronic nose based on metal oxide semiconductor sensors as a fast alternative for the detection of adulteration of virgin olive oils. Analytica Chimica Acta, 459, 219–228.

[110] Cosio, M. S., Ballabio, D., Benedetti, S., and Gigliotti, C. (2006) Geographical origin and authentication of extra virgin olive oils by an electronic nose in combination with artificial neural networks. Analytica Chimica Acta, 567, 202–210.

[111] Hai, Z. and Wang, J. (2006) Electronic nose and data analysis for detection of maize oil adulteration in sesame oil. Sensors and Actuators. B, Chemical, 119, 449–455.
[112] Pioggia, G., Ferro, M., and Di Francesco, F. (2007) Towards a real-time transduction and classification of chemoresistive sensor array signals. IEEE Sensors Journal, 7, 227–241.
[113] Cosio, M. S., Ballabio, D., Benedetti, S., and Gigliotti, C. (2006) Geographical origin and authentication of extra virgin olive oils by an electronic nose in combination with artificial neural networks. Analytica Chimica Acta, 567, 202–210.
[114] Zhang, Q., Zhang, S., Xie, C., Fan, C., and Bai, Z. (2008) Sensory analysis of Chinese vinegars using an electronic nose. Sensors and Actuators. B, Chemical, 128, 586–593.
[115] Brezmes, J., Ferreras, B., Llobet, E., Vilanova, X., and Correig, X. (1997) Neural network based electronic nose for the classification of aromatic species. Analytica Chimica Acta, 348, 503–509.
[116] Schaller, E., Bosset, J. O., and Escher, F. (1998) Electronic noses and their application to food. Food Science & Technology, 31, 305–316.
[117] Gardner, J. W., Shurmer, H. V., and Tan, T. T. (1992) Application of an electronic nose to the discrimination of coffees. Sensors and Actuators. B, Chemical, 6, 71–75.
[118] Bartlett, P. N., Blair, N., and Gardner, J. W. (1993) Electronic nose. Principles, applications and outlook, in ASIC, 15th colloque, Montpellier, France; Paris, France: ASIC, pp. 478–486.
[119] Fukunaga, T., Mori, S., Nakabayashi, Y., Kanda, M., and Ehara, K. (1995) Application of flavor sensor to coffee, in ASIC, 16th International Scientific Collopuium on Coffee, Kyoto, Japan; Paris, France: ASIC, pp. 478–486.
[120] Falasconi, M., Pardo, M., Sberveglieri, G., Ricco, I., and Bresciani, A. (2005) The novel EOS835 electronic nose and data analysis for evaluating coffee ripening. Sensors and Actuators. B, Chemical, 110, 73–80.
[121] Dias, L. A., Peres, A. M., Vilas-Boas, M., Rocha, M. A., Estevinho, L., and Machado, A. A. S. C. (2008) An electronic tongue for honey classification. Mikrochimica Acta, 163, 97–102.
[122] Escriche, I., Kadar, M., Domenech, E., and Gil-Sánchez, L. (2012) A potentiometric electronic tongue for the discrimination of honey according to the botanical origin. Comparison with traditional methodologies: physicochemical parameters and volatile profile. Journal of Food Engineering, 109, 449–456.
[123] Wei, Z., Wang, J., and Liao, W. (2009) Technique potential for classification of honey by electronic tongue. Journal of Food Engineering, 94, 260–266.
[124] Major, N., Marković, K., Krpan, M., Šarić, G., Hruškar, M., and Vahčić, N. (2011) Rapid honey characterization and botanical classification by an electronic tongue. Talanta, 85, 569–574.
[125] Legin, A., Rudnitskaya, A., Seleznev, B., and Vlasov, Y. (2002) Recognition of liquid and flesh food using an 'electronic tongue'. International Journal of Food Science & Technology, 37, 375–385.
[126] Legin, A., Rudnitskaya, A., Vlasov, Y., Di Natale, C., Davide, F., and D'Amico, A. (1997) Tasting of beverages using an electronic tongue. Sensors and Actuators. B, Chemical, 44, 291–296.
[127] Ciosek, P., Brzózka, Z., and Wróblewski, W. (2004) Classification of beverages using a reduced sensor array. Sensors and Actuators. B, Chemical, 103, 76–83.
[128] Legin, A., Rudnitskaya, A., Vlasov, Y., Di Natale, C., Mazzone, E., and D'Amico, A. (1999) Application of electronic tongue for quantitative analysis of mineral water and wine. Electroanalysis, 11, 814–820.
[129] Gallardo, J., Alegret, S., and del Valle, M. (2005) Application of a potentiometric electronic tongue as a classification tool in food analysis. Talanta, 66, 1303–1309.
[130] Ciosek, P., Augustyniak, E., and Wróblewski, W. (2004) Polymeric membrane ion-selective and cross-sensitive electrodes-based electronic tongue for qualitative analysis of beverages. The Analyst, 129, 639–644.

[131] Ciosek, P., Maminśka, R., Dybko, A., and Wróblewski, W. (2007) Potentiometric electronic tongue based on integrated array of microelectrodes. Sensors and Actuators. B, Chemical, 127, 8–14.

[132] Salter, G. J., Lazzari, M., Giansante, L., Goodacre, R., Jones, A., Surricchio, G., Kell, D. B., and Bianchi, G. (1997) Determination of the geographical origin of Italian extra virgin olive oil using pyrolysis mass spectrometry and artificial neural networks. Journal of Analytical and Applied Pyrolysis, 40–41, 159–170.

[133] Kocjancic, R. and Zupan, J. (1997) Application of a feed-forward artificial neural network as a mapping device. Journal of Chemical Information and Modeling 37, 9859.

[134] Bertran, E., Blanco, M., Coello, J., Iturriaga, H., Maspoch, S., and Montoliu, I. (2000) Near infrared spectrometry and pattern recognition as screening methods for the authentication of virgin olive oils of very close geographical origins. Journal of Near Infrared Spectroscopy, 8, 45–52.

[135] Rezzi, S., Axelson, D. E., Héberger, K., Reniero, F., Mariani, C., and Guillou, C. (2005) Classification of olive oils using high throughput flow H NMR fingerprinting with principal component analysis, linear discriminant analysis and probabilistic neural networks. Analytica Chimica Acta 552, 13–24.

[136] García-González, D. L., Luna, G., Morales, M. T., and Aparicio, R. (2009) Stepwise geographical traceability of virgin olive oils by chemical profiles using artificial neural network models. European Journal of Lipid Science and Technology 111(10), 03–13.

[137] García-Gonzélez, D. L. and Aparicio, R. 2003. Virgin olive oil quality classification combining neural network and MOS sensors. Journal of Agricultural and Food Chemistry 51(12), 3515–3519.

[138] Cajka, T., Riddellova, K., Klimankova, E., Cerna, M., Pudil, F., and Hajslova, J. (2010) Traceability of olive oil based on volatiles pattern and multivariate analysis. Food Chemistry, 121(1), 282–289.

[139] Mannina, L. and Sobolev, A. P. (2011) High resolution NMR characterization of olive oils in terms of quality, authenticity and geographical origin. Magnetic Resonance in Chemistry, 49, S3–11.

[140] Caceres, J. O., Moncayo, S., Rosales, J. D., Javier, F., De Villena, M., Alvira, F. C., and Bilmes, G. M. 2013. Application of laser – induced breakdown spectroscopy (LIBS) and neural networks to olive oils analysis. Applied Spectroscopy 67, 1064–1072.

[141] Torrecilla, J. S., Cancilla, J. C., Matute, G., and Díaz-Rodríguez, P. (2013a) Neural network models to classify olive oils within the protected denomination of origin framework. International Journal of Food Science & Technology 48, 2528–2534.

[142] Torrecilla, J. S., Cancilla, J. C., Matute, G., Díaz-Rodríguez, P., and Flores, A. I. (2013b) Self-organizing maps based on chaotic parameters to detect adulterations of extra virgin olive oil with inferior edible oils. Journal of Food Engineering 11(8), 400–405.

[143] Gonzalez Martin, Y., Concepcion, M., Oliveros, C., Luis, J., Pavón, P., Pinto, C. G., and Cordero, B. M. (2001) Electronic nose based on metal oxide semiconductor sensors and pattern recognition techniques: characterisation of vegetable oils. Analytica Chimica Acta 44(9), 69–80.

[144] Concepcion, M. A., Oliveros, C., Luis, J., Pavón, P., Pinto, C. G., Laespada, E. F., Moreno Cordero, B., and Forina, M. (2002) Electronic nose based on metal oxide semiconductor sensors as a fast alternative for the detection of adulteration of virgin olive oils. Analytica Chimica Acta 45(9), 219–228.

[145] Ali, Z., James, D., O'Hare, W. T., Rowell, F. J., and Scott, S. M. (2003) Radial basis neural network for the classification of fresh edible oils using an electronic nose. Journal of Thermal Analysis and Calorimetry, 71, 147–154.

[146] García-Gonzélez, D. L., Luna, G., Morales, M. T., and Aparicio, R. (2009) Stepwise geographical traceability of virgin olive oils by chemical profiles using artificial neural network models. European Journal of Lipid Science and Technology 11(1), 1003–1013.
[147] Torrecilla, J. S., Mena, M. L., Yáñez-Sedēno, P., and García, J. (2007) Quantification of phenolic compounds in olive oil mill wastewater by artificial neural network/laccase biosensor. Journal of Agricultural and Food Chemistry 55, 7418–7426.
[148] Jiménez, A., Beltran, G., Aguilera, M. P., and Uceda, M. (2008) A sensor-software based on artificial neural network for the optimization of olive oil elaboration process. Sensors and Actuators. B, Chemical. 12(9), 985–990.
[149] Kussul, N., Lavreniuk, M., Skakun, S., and Shelestov, A. (2017) Deep learning classification of land cover and crop types using remote sensing data. IEEE Geoscience and Remote Sensing Letters 14(5), 778–782.
[150] Herranz, L., Jiang, S., and Xu, R. (2017) Modeling restaurant context for food recognition. IEEE Transactions on Multimedia, 19(2), 430–440.

3 Multi-layered perceptron

3.1 Artificial neural networks: principles and training algorithms

The demonstrated inability of a single Perceptron, trained with the perceptron, Adaline, or the Delta learning rules, to resolve interesting problems resulted in a notable research slowdown on connectionism; see Chapter 1. However, some work demonstrated in the course of the 1970s that one could still overcome these limitations by interposing between the input layer and the output layer, one or more intermediate layers (also called hidden layers), information flowing from one layer to the next given the well-known feedforward network (Figure 3.1).

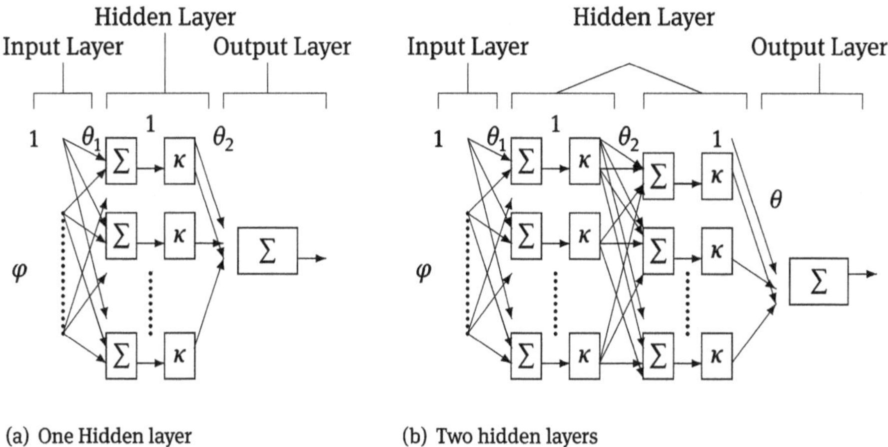

(a) One Hidden layer

(b) Two hidden layers

Figure 3.1: Neurons organised in network structure.

In this case, a solution for nonlinearly separable classification problems, as well as complex nonlinear approximation function one is nearly guaranteed. In the next section, the Cybenko theorem will be explained as a proof of the ability of Multi-layered Perceptron to approximate any nonlinear function. However, and for now the notion of network needs to be explained and presented:

where φ is the input vector (called regressor) and θ is the neuron's activation functions used; i. e., sigmoidal functions in the case of the MLP.

3.2 Regressor type and network topologies

Largely inspired by linear systems theory, [1] and [5], a number of regressors were selected in order to be used in the nonlinear case, and in particular with MLPs [2, 6, 3].

https://doi.org/10.1515/9783110646054-003

This is convenient because as established for the single neuron, the nonlinearity only appears in the connection of the regressor at the outputs. If the estimated output is $\hat{y}$ at time k, it is given by

$$\hat{y}(k|\theta) = g(\varphi(k), \theta) \tag{3.1}$$

Note that previous values of $\hat{y}$ are function of $g(\varphi(k))$ and, therefore, function of θ. It is thus the reason why $\hat{y}$ at time k is given by $\hat{y}(k|\theta)$. It is then possible to distinguish the following regressors:

- Finite impulse response structure:

$$\varphi = \left[u(k-d)\ \ u(k-1-d) \cdots u(k-m-d)\right]^T \tag{3.2}$$

- Autoregressive with exogenous inputs model structure (NARX):

$$\varphi = \left[y(k-1)\ \ y(k-2) \cdots y(k-n)\ \ u(k-1-d)\ \ u(k-2-d) \cdots u(k-m-d)\right]^T \tag{3.3}$$

- Output error model structure (NOE):

$$\varphi = \left[y(k-1|\theta)\ \ y(k-2|\theta) \cdots y(k-n|\theta)\ \ u(k-1-d)\ \ u(k-2-d) \cdots u(k-m-d)\right]^T \tag{3.4}$$

- Auto-regressive moving average model structure (ARMAX):

$$\begin{aligned}\varphi = \big[&y(k-1)\ \ y(k-2) \cdots y(k-n)\ \ u(k-1-d)\\ &u(k-2-d) \cdots u(k-m-d)\ \ \epsilon(k-1) \cdots\ \ \epsilon(k-r)\big]^T\end{aligned} \tag{3.5}$$

  where, $\epsilon(k) = y(k) - \hat{y}(k|\theta)$.
- Box–Jenkins structure (NBJ):

$$\begin{aligned}\varphi = \big[&y(k-1|\theta)\ \ y(k-2|\theta)\ \ \cdots\ \ y(k-n|\theta)\ \ u(k-1-d)\ \ u(k-2-d) \cdots\\ &u(k-m-d)\ \ \epsilon(k-1) \cdots\ \ \epsilon(k-r)\ \ \epsilon_u(k-1) \cdots\ \ \epsilon_u(k-r)\big]^T\end{aligned} \tag{3.6}$$

  where $\epsilon_u(k) = y(k) - \hat{y}_u(k|\theta)$, $\hat{y}_u(k|\theta)$ are the simulated outputs using only passed values of u.

Note that many other ANN topologies organised as networks exists, and used for different application fields. The most important ones will be discussed in details in Chapters 4, 5, and 6 for respectively RBFs, Kohonen maps and deep networks architectures.

3.3 Multi-layer perceptron topology

A Multi-Layered feedforward neural network is a set of interconnected processing units consisting in one input layer, one or more hidden layers and one output layer. This shape allows the captured information injected in the input layer to be transferred through hidden layers until it reaches the output layer in one directional way only (no feedback loops). The network parameters are then adjusted with a repeated disclosure to input/output training patterns. During the training process, the network error is obtained by calculating the difference between the network output and the desired response, the network parameters are then repeatedly adjusted until the optimum state is reached where the network output becomes sensibly close (in a predefined error range) to the supervised output.

Basically, there are two types of feedforward architectures: single hidden layer structure and two hidden layers structure; see Figure 3.1. In both, the activity of the input neurons is restricted to representing unprocessed information only, whereas the role of the hidden neurons in the case of one or two layers network is affected by the input signal and the weight's connections between input and hidden neurons. The output reaction is therefore relying on the activations of the hidden neurons multiplied with the hidden-output connections weights. As described in Section 1.7, the Minsky and Papert's limitation pointed that in order to handle a nonlinear tasks, the multilayer architecture networks would definitely provide more computational efficiency compared with the single-layer structure networks, which was practically proven on the XOR problem.

Technically, Figure 3.2 illustrates a Multi-Layered Perceptron (MLP) network with D input units, C output units, and several hidden units, which all are arranged in layers. The jth units in layer l computes the output according the following formula:

$$y_j = \theta(z_j) \tag{3.7}$$

where

$$z_j = \sum_{i=1}^{K} w_{ji} y_i + w_{j0} \tag{3.8}$$

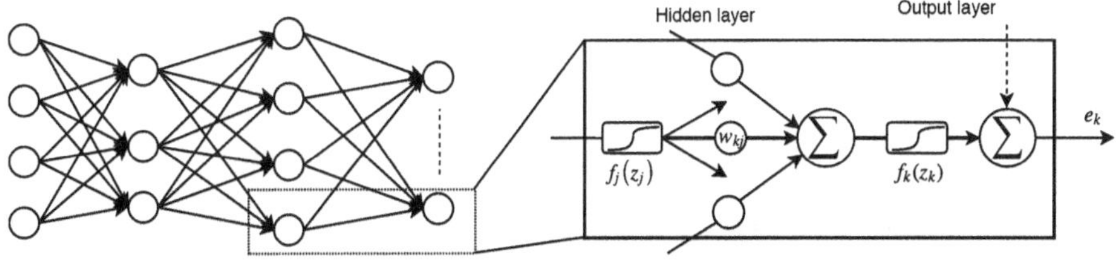

Figure 3.2: The MLP network structure with information processing through neurons.

where w_n is the weighted connection between the nth neuron to the ith neuron in the next layer, and $w_{i,0}$ corresponds to the bias which considered as an external input to the neuron. Here, K denotes the number of neurons in the source layer. The θ denotes the activation function.

3.4 Activation function

In order to compute and limit the amplitude range of the output y_j for each neuron of the multi-layer perceptron, a continuously differentiable activation function is required to be associated with the neurons, Figure 3.3. Basically, there are two identified types of activation function namely, threshold functions and sigmoidal functions.

1. Threshold function: The output in this type as illustrated in Figure. 3.4(a) is given by

$$\text{Threshold}(z_j) = \begin{cases} 1 & \text{if } z_j \geq 0 \\ 0 & \text{if } z_j > 0 \end{cases}$$

2. Sigmoidal nonlinearity: Unlike threshold units, sigmoidal units are so sensitive when the sum of inputs z_j is near 0. Consequently, these types of functions are strongly recommended for constructing a multi-layered perceptrons. There are two forms of the sigmoidal function to identify, Figure 3.4(b):
   (a) Logistic function: An example of the sigmoidal functions taking a S-shape, transforming the sum of input values into a range between [0,1].

$$\sigma(x) = \frac{1}{1 + e^{-z_j}} \tag{3.9}$$

   (b) Hyperbolic tangent function: This activation function also has a S-shape where the output value lies in the rage of $0 \leq y_j \leq 1$. According to this function, it saturates to −1 or 1 when z_j becomes highly negative or highly positive,

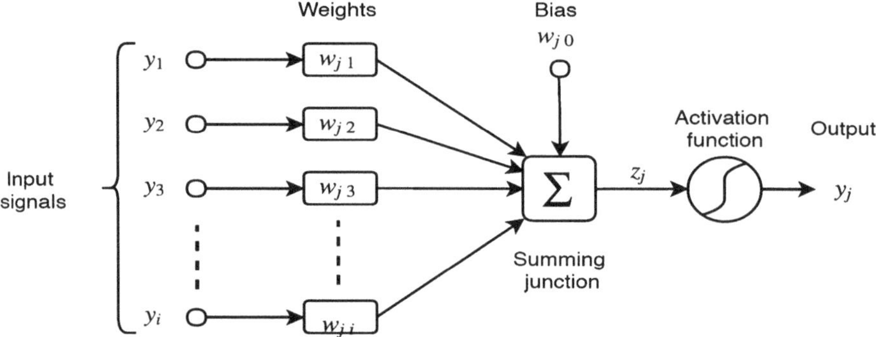

Figure 3.3: Illustration of a nonlinear neuron, labelled j.

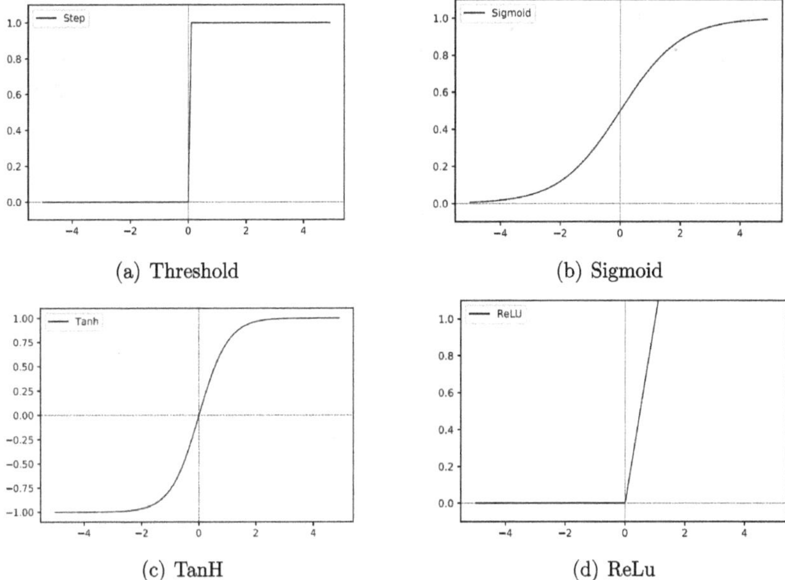

(a) Threshold (b) Sigmoid

(c) TanH (d) ReLu

Figure 3.4: Activation functions.

respectively, Figure 3.4(c).

$$\tanh(z_j) = \frac{2}{1 + e^{-2z_j}} - 1 \tag{3.10}$$

- In addition to the above activation functions, one may cite the less frequently used, such as the Rectified (ReLU) activation function (see Figure 3.4(d)), also known as the rectified linear unit. A very simple function, whereas each value's amplitude $z_j < 0$ is truncated at 0 and keep it as it is if $z_j > 0$ as expressed below:

$$\text{ReLU}(z_j) = \begin{cases} 0 & \text{if } z_j < 0 \\ z_j & \text{otherwise} \end{cases} \tag{3.11}$$

The output amplitude can therefore take many forms. Theoretically, there is no defined strategy to find the correct activation function that should be used for a specific condition, in most cases, an empirical process is recommended in order to properly select the appropriate activation function for a given task.

3.5 Cybenko theorem

According to Cybenko's theorem (1989) [7] and its extended proof in [12], which states the following:

Let $f(.)$ be a continuous, bounded, and monotonic function. Let I_p gives the hypercube $[0,1]^p$ of dimension p. The space of the continuous function on I_p is given by $C(I_p)$. So if $f \in C(I_p)$ and $\varepsilon > 0$, there is an integer M and a number of real constants α_i, β_i, and w_{ij} where $i = 1, \ldots, M$ such that

$$\widehat{F}(x_1, \ldots, x_p) = \sum_{i=1}^{M} \alpha_i f\left(\sum_{j=1}^{p} w_{ij} x_j - \beta_i\right) \tag{3.12}$$

is seen as an approximation of function $F(.)$, where

$$\left|F(x_1, \ldots, x_p) - \widehat{F}(x_1, \ldots, x_p)\right| < \varepsilon \tag{3.13}$$

for all $x_1, \ldots, x_p$.

Figure 3.5 shows a feedforward MLP with only one output F, according to equation (3.12) and constructed according to Cybenko's theorem.

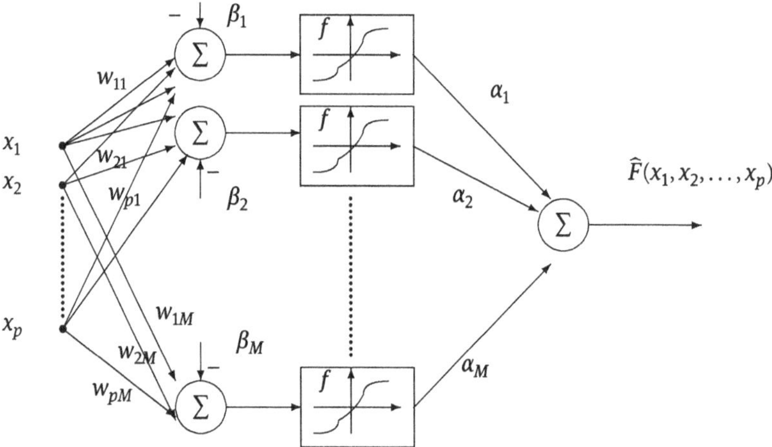

Figure 3.5: MLP topology justified by the Cybenko theorem.

Note that:

- Tan-sigmoid and log-sigmoid are continuous, bounded, and monotonous functions, respecting Cybenko's assumptions.
- Equation (3.12) describes a number of entries p, and a single hidden layer of M neurons.
- The hidden neuron i has the synaptic weights $w_{i1}, w_{i2}, \ldots, w_{ip}$ and the biases β_i.
- Network output is a linear combination of outputs corresponding to the hidden neurons with the weights $\alpha_1, \alpha_2, \ldots, \alpha_M$.
- Cybenko's theorem is an existence theorem, so it does not specify the number of neurons in the hidden layer.

- In practice, and as the theorem remains true for many hidden layers of neurons, it could be more profitable to use more than one hidden layer.

In conclusion, Cybenko describes a single layer MLP, and states that giving an appropriate number of neurons and adequate weights and biases, the network can approximate any nonlinear function. Theoretically, this is a very important assumption, solving eventually all nonlinear problems. However, the Cybenko theorem is only an existence one, and it only states the existence of a "particular" MLP topology as a solution for any nonlinear problem, not stating how to find the network parameters: number of inputs, number of layers, number of neurons, training approach, etc.

As the theorem stands for multiple hidden layers, the "primary" question that strikes is: shall we use one, two or more hidden layers?

In the case of a single-layered MLP, the neurons interact in a global way (as a global classifier). Therefore, when approximating a complex function, an improvement in part of the input space, may cause deterioration in another part (global approximation). On the other hand, in the case of a network with two hidden layers the following consequences are noticed:

- Local characteristics are extracted in the first hidden layer. In other words, the first layer partition the input space into regions, there follows a local characteristics learning.
- The global characteristics are then extracted in the second hidden layer, where neurons of the latter combines the outputs of the neurons of the first layer, which operate around an input space region building the overall characteristics of this region but producing zeros for the rest of the input space.
- In addition, certain problems (using discontinuous neurons types) cannot be resolved with a single hidden layer, but can be solved by a two hidden layer networks.

Since the introduction of deep learning, the questioning of multiple layers is no longer an issue. The computational power offered nowadays and Deep learning (DL) training approaches, in addition to the high dimensionality problems and abundance of data, favors the usage of large numbers of layers and neurons.

3.6 Network training

3.6.1 Supervised learning

Considering learning with a teacher (or an observer) refers to the supervised learning paradigm, Figure 3.6. The basic concept of neural network supervised training is to be guided by an external knowledge from the environment, represented by a set of input-output samples. These built-in knowledge samples are then used as training

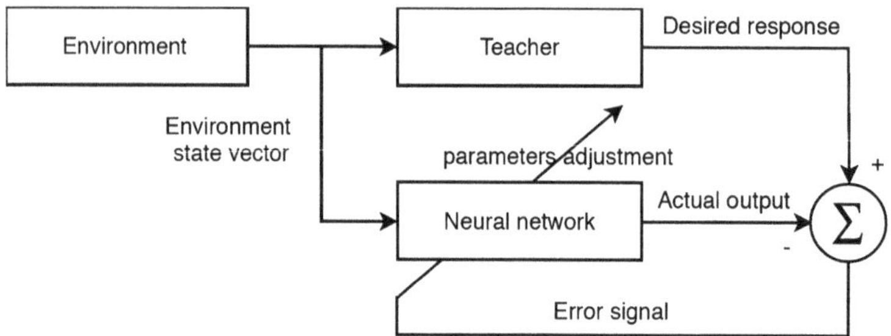

Figure 3.6: Illustration of the supervised learning paradigm.

vectors for the neural network providing the optimum desired actions or responses to be performed in order to match the data behaviour or interactions. The neural network parameters are then updated according to both influences of the input vector and the produced error signal, which is the difference between the desired response and the actual the network output. After adjusting the parameters in an iterative manner, the emulation of the network behaviour (with the help of the teacher, or observer) is presumed. Thus once the neural network becomes converge to a suitable solution, and all the available knowledge from the environment are properly learned, the network is left to deal with the environment completely by itself.

3.6.2 Unsupervised learning

For unsupervised learning, also called self-organised learning, there is no external observer or teacher to oversee and monitor the training process. Thus there is no need for desired targets to be relied on. Instead, as indicated in Figure. 3.7, the training is mainly concerned in finding the appropriate network parameters with respect to the similarities in the data, using only available nonlabelled inputs. After optimizing the parameters according to a specific task-independent measure, the network will be able to create new classes based on an internal representation of the input data.

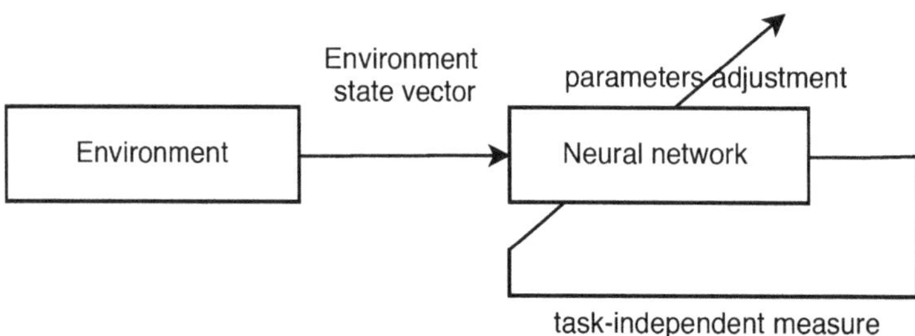

Figure 3.7: Illustration of the unsupervised learning paradigm.

3.6.3 Back-propagation learning

Back-Propagation (BP) learning counts among supervised learning techniques (Figure 3.6), where the output layer considers only one neuron, where a neuron k is the only computational unit. The learning rule is based on the single neural network output signal, $y_k(n)$ and its comparison with the desired response or the target output $d_k(n)$, in order to compute the error $e_k(n)$ at each nth time step according the following definition:

$$e_k(n) = d_k(n) - y_k(n) \tag{3.14}$$

In an iterative process of adjusting the neuron k's synaptic weights and biases, the corrective adjustment is applied to minimise the gap between the network output and the desired response. The process adopts a cost function also called an index of performance based on the error $E(n)$ instead of using the standard error $e_k(n)$ defined as

$$E(n) = e_k^2(n) \tag{3.15}$$

The search is based on gradient search, where the simplest method, known as steepest descent or gradient descent, which was originally considered by [13], is to find the derivative of the cost function with respect to the network weights, and then update the weights by taking a small fixed-size step in the direction of the negative error gradient of the loss function according to the formula or the update rule:

$$w_{ji}(t+1) = w_{ji}(t) + \Delta w_{ji}(t) \tag{3.16}$$

and

$$\Delta w_{ji}(t) = -\mu \frac{\partial E}{\partial w_{ji}} \tag{3.17}$$

where ($\mu \in [0,1]$) is the learning rate. This computation is repeated to reach some stopping criteria such as falling into a local minima where it fails to reduce the loss or accomplishing a given number of steps.

Local minima phenomena is considered as one of the most crucial problems that faces the gradient descent method while updating the weights. To avoid getting trapped in such a minima, [14] had introduced a momentum factor m that effectively allows the network to ignore the insignificant features in the error and, therefore, improve the selected final state of the network:

$$\Delta w_{ji}(t+1) = -\mu \frac{\partial E}{\partial w_{ji}} + \beta \Delta w_{ji}(t) \tag{3.18}$$

$\beta \in [0,1]$ also chosen in the range of [0,1].

The gradients could be calculated and defined over the entire training dataset (batch learning), over small subsets (sequential learning), or over every individual training example (online learning) where the latest tends to be more efficient then both sequential and batch learning in the case of dealing with big amount of data with significant redundancy [15]. In the same aspect, stochasticity in online learning according to [15], also can help prevent the network from falling into the local minima due to the diversity of the loss function for each training sample by randomly organizing these samples before using it in the training process. Another stochastic online learning that was proposed by [19] is the stochastic meta-descent that shows a fast convergence and considerable improved results for different tasks.

The BP algorithm, using the *SE* (equation (3.15)), is based on two distinct stages: a forward pass and a backward pass. In the first stage, the final network error is calculated by obtaining the network output, propagating the input vector through the network layers and parameters (weight and biases). Once the network output is calculated, the error between simulated and real output values is obtained and used in the back-propagation stage, updating each neuron weight and biases, backward from the output layer to the first hidden layer. In summary, the forward pass propagates inputs through the network without the weights' update to calculate the network error. Meanwhile the backward pass uses the calculated error and the delta learning rule to update neuron weights.

3.6.4 Error function derivatives

Now considering the evaluation of the $E(n)$ regarding the weight w_{ji} related to unit j, where the output of each neuron depends on the particular input pattern. In order to find the chain rule for a partial derivative, it is essential to consider z_j due to the inclusion of w_{ji} as formulated in equation (3.8). Hence, we note

$$\frac{\partial E}{\partial w_{ji}} = \frac{\partial E}{\partial z_j}\frac{\partial z_j}{\partial w_{ji}} \tag{3.19}$$

Now we introduce

$$\delta_j = \frac{\partial E}{\partial z_j} \tag{3.20}$$

using the activations received from the neurons in the previous layer i, we can write

$$\frac{\partial z_j}{\partial w_{ji}} = y_i \tag{3.21}$$

Substituting equations (3.20) and (3.21) into (3.19), we then obtain

$$\frac{\partial E}{\partial z_j} = \delta_j y_i \tag{3.22}$$

3.6.5 Update process

Let us begin by reminding the Delta learning rule seen in Section 1.10.2 and given here in equation (3.23).

$$W_{k+1} = W_k + 2\mu e f'(z) X \tag{3.23}$$

On the one hand, and for each output neuron, there is no problem of implementing the Delta rule as the error ($e = d - y$) is directly available. On the other hand, for the hidden layer's neurons (if we consider a network with a single hidden layer), we are in the obligation to retro propagate the output error in order to obtain an effective error at the output of each neuron from the hidden layer, Figure 3.8. This is developed mathematically in what follows.

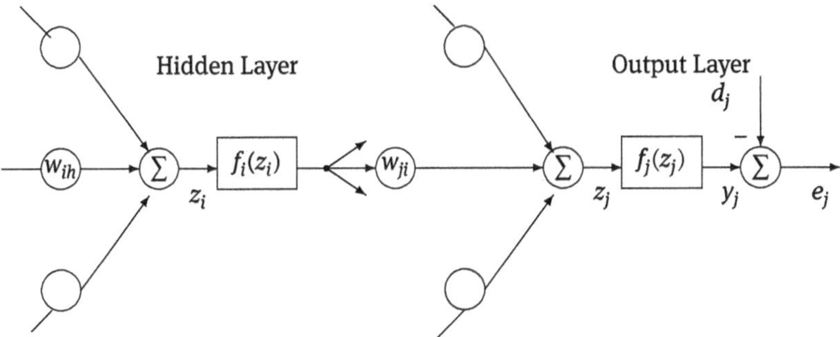

Figure 3.8: Two neurons belonging to different layers in a one hidden layer MLP.

First, let us consider a neuron j in the output layer:

The cost function is then given by

$$J = \sum_{j=1}^{N} e_j^2 \tag{3.24}$$

$$J = \sum_{j=1}^{N} J_j \tag{3.25}$$

where each J_j depends only on neuron j. Minimisation of each J_j independently from the rest becomes possible.

The gradient of neuron j in the output layer is given by

$$\frac{\partial J}{\partial w_{ji}} = \frac{\partial J \partial e_j \partial z_j}{\partial e_j \partial z_j \partial w_{ji}} \tag{3.26}$$

We have then

$$\frac{\partial J}{\partial e_j} = 2e_j \tag{3.27}$$

$$\frac{\partial e_j}{\partial z_j} = -f_j'(z_j) \tag{3.28}$$

$$\frac{\partial z_j}{\partial w_{ji}} = y_i \tag{3.29}$$

defining, δ_j as the local gradient through neuron j, as follows:

$$\delta_j = -\frac{\partial J}{\partial z_j} = 2e_j f_j'(z_j) \tag{3.30}$$

thus,

$$\frac{\partial J}{\partial w_{ji}} = -2e_j f_j'(z_j) y_i = -\delta_j y_i \tag{3.31}$$

The update rule becomes then

$$\Delta w_{ji} = -\mu \frac{\partial J}{\partial w_{ji}} = 2\mu e_j f_j'(z_j) y_i = \mu \delta_j y_i \tag{3.32}$$

If neuron j is in a hidden layer (Figure 3.9), then the cost function becomes

$$J = \sum_{k=1}^{M} e_k^2 \tag{3.33}$$

It is needed to determine the partial gradient $\frac{\partial J}{\partial w_{ji}}$ for updating the weights w_{ji}, noting that w_{ji} will affect all output errors $e_1, e_2, \ldots, e_M$:

$$\frac{\partial J}{\partial w_{ji}} = \frac{\partial J \partial z_j}{\partial z_j \partial w_{ji}} \tag{3.34}$$

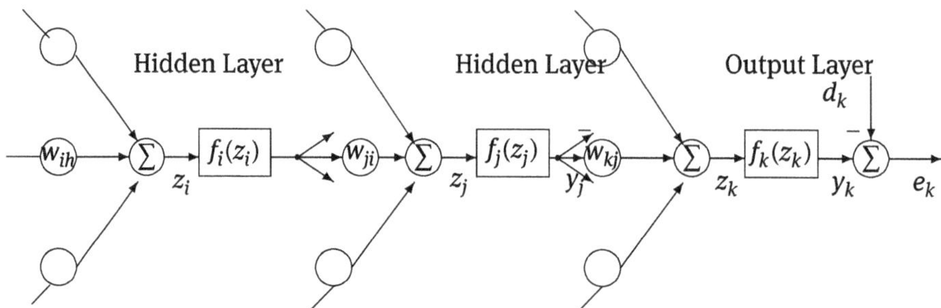

Figure 3.9: Three neurons in different layers of a two hidden layer MLP.

where

$$\frac{\partial z_j}{\partial w_{ji}} = y_i \tag{3.35}$$

and,

$$\delta_j = \frac{\partial J}{\partial z_j} = \frac{\partial J \partial y_j}{\partial y_j \partial z_j} \tag{3.36}$$

with

$$\frac{\partial J}{\partial z_j} = f_j'(z_j) \tag{3.37}$$

From (3.33), we have

$$\frac{\partial J}{\partial y_j} = 2\sum_{k=1}^{M} e_k \frac{\partial e_k}{\partial y_j} \tag{3.38}$$

obtaining

$$\frac{\partial e_k}{\partial y_j} = \frac{\partial e_k \partial z_k}{\partial z_k \partial y_j} \tag{3.39}$$

with

$$\frac{\partial e_k}{\partial z_k} = -f_k'(z_k) \quad \text{and} \quad \frac{\partial z_k}{\partial y_j} = w_{kj} \tag{3.40}$$

combining(3.37) and (3.40), we obtain

$$\frac{\partial J}{\partial y_j} = -2\sum_{k=1}^{M} e_k f_k'(z_k) w_{kj} = -2\sum_{k=1}^{M} \delta_k w_{kj} \tag{3.41}$$

From equation (3.30), knowing that neuron k is an output neuron, we obtain

$$\delta_j = -\frac{\partial J}{\partial z_j} = \frac{\partial J \partial y_j}{\partial y_j \partial z_j} = 2f_j'(z_j)\sum_{k=1}^{M} \delta_k w_{kj} \tag{3.42}$$

Finally, the update rule is given by

$$\Delta w_{ji} = -\mu \frac{\partial J}{\partial w_{ji}} = \mu \delta_j y_i \tag{3.43}$$

$$\Delta w_{ji} = 2\mu f_j'(z_j)\frac{\partial J}{\partial w_{ji}} = \mu f_j'(z_j)\sum_{k=1}^{M} \delta_k w_{kj} y_i \tag{3.44}$$

In summary,

$$\Delta w_{ji} = \mu \delta_j y_i \tag{3.45}$$

- If j is in an output layer, then $\delta_j = 2f_j'(z_j)e_j$.
- If j is in a hidden layer, then $\delta_j = 2f_j'(z_j)\sum_{k=1}^{M} \delta_k w_{kj}$.

3.7 Exclusive OR (XOR) classification using MLP

3.7.1 Analytical solution

As stated earlier in Section 1.7, a single neuron is not able to classify the nonlinearly separable XOR function, and at least two neurons plus a combination one are needed. The problem is revisited here using both an analytical graphical solution and an iterative one using the BP algorithm.

For a graphical analytical solution (Figures 3.11 and 3.12), the decisions made by neuron 1 and neuron 2 in Figure 3.10 , is given by the following linear equations, respectively:

$$x_2 > -x_1 + 1.5 \quad (y_1 = 1)$$

and

$$x_2 > x_1 + 0.5 \quad (y_2 = 1)$$

The output neuron implements then the combination of y_1 and y_2 in the following equation:

$$-2y_1 + y_2 > 0.5 \equiv [\text{NOT}(y_1)] \quad \text{AND} \quad y_2$$

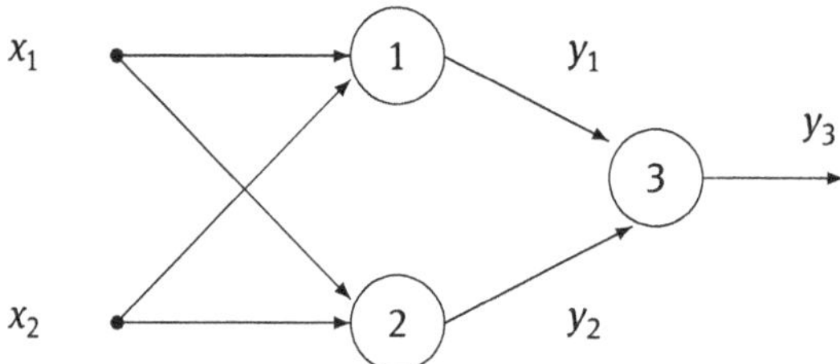

Figure 3.10: Three neurons MLP.

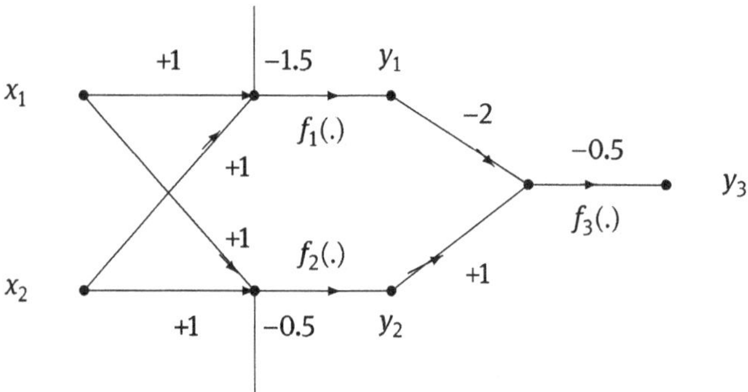

Figure 3.11: Weights and biases of the MLP.

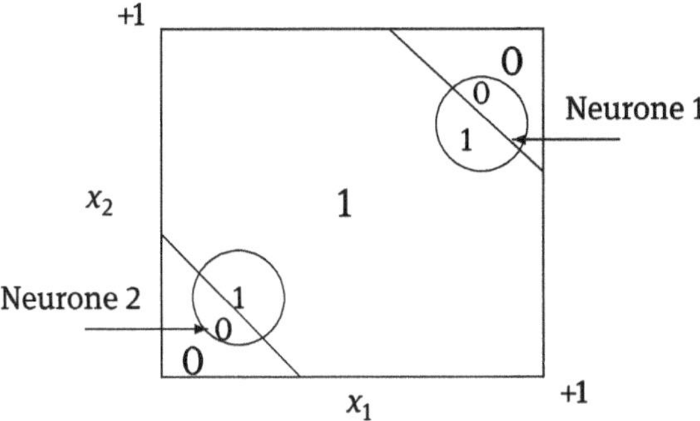

Figure 3.12: MLP decision diagram.

3.7.2 Adaptive solution

In what follows, the XOR function is implemented in Matlab using the function *newp* to create the MLP network and the function *train* for training the MLP. The implementation details are given by:

- 2 inputs + 2 sigmoidal neurons + 1 linear neuron.
- Training for 15 epochs with a learning rate $\mu = 0.1$.
- Random initialisation of weights and biases.
- $P = \begin{bmatrix} 0 & 1 & 0 & 1 \\ 0 & 0 & 1 & 1 \end{bmatrix} = \begin{bmatrix} x_1 \\ x_2 \end{bmatrix}$, $T = [0 \ 1 \ 1 \ 0]$.

Performances after 5 epochs:

$$T_{\text{test}} = [0.0000 \quad 1.0000 \quad 1.0006 \quad 0.0002]$$

We can see that a MLP with 1 hidden layer, containing 2 neurons, solves the XOR problem easily, where the perceptron alone cannot reach a solution as it is enable to solve nonseparably linear problems. Learning is very fast, as the problem is relatively simple, and after 3 epochs of training (Figure 3.13), the reference error of 0.0001 is reached.

3.8 Enhancing MLP's training

MLP training may be enhanced using many existing approaches and methods, in order to make it:

- more robust
- less sensitive to local minimums
- faster in terms of convergence time.

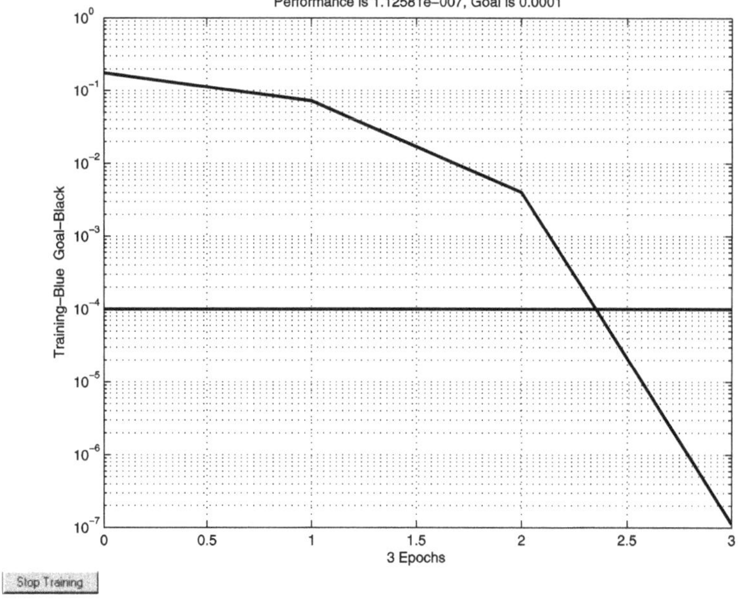

Figure 3.13: MLP training to solve the XOR problem.

In what follows, a couple of approaches are cited and described. Note that this is not an extensive study, and may just show the reader how training enhancement may be performed. Examples and a wider explanation on the enhancement of the back-propagation algorithm is given in [8, 9].

Nowadays most of learning libraries in different languages offer a large scope of BP versions containing training enhancements with regards to the original algorithms. The users have then all the latitude to test and choose the appropriate ones.

3.8.1 Inertia moment approach

This approach consists in adding an inertia moment to the weight and biases variation during training. This has three main advantages:

- Speeding convergence.
- Increasing chances to avoid local minimums.
- Help smoothing weights evolution through filtering.

The gradient decent update rule is then modified as follows:

$$\Delta w_k = -\mu J_k + \alpha \Delta w_{k-1} \tag{3.46}$$

For a total of N iterations while using an inertia moment, the weights update may be expressed by:

$$\Delta w_k = -\mu \Sigma_{n=0}^{N} \alpha^n \nabla J_{k-1} \tag{3.47}$$

3.8.2 Adaptive learning step

In this modification of the BP algorithm, the learning step is progressively augmented if the cost function (Sum Squared Error (SSE), quadratic error, etc.) keeps decreasing. It is on the contrary strongly decreased in the case of an increasing cost function.

An example of the aforementioned adaptive approach is given by

$$\Delta\mu = \begin{cases} +a & \text{If } \Delta J < 0, \\ -b\mu & \text{If } \Delta J > 0, \\ 0 & \text{Else.} \end{cases} \tag{3.48}$$

A substantial enhancing would be to undo the results of a "bad iteration" respecting the condition equation (3.47).

3.8.3 Weight update using new entries

This approach describes a weight update based on new entries, i. e., a weight is updated each time that a new input vector is presented to the network. An alternative is to update the weights after each group of input vector, the gradient used is therefore an average of gradients for each input vector and, therefore, corresponds more to the true Mean Squared Error (SSE) gradient. From where, minimisation of the following cost function:

$$J_{\text{moy}} = \frac{1}{P} \sum_{p=1}^{P} \sum_{i=1}^{N} e_i^2(p) \tag{3.49}$$

This approach, on the other hand, requires more storage memory and is more sensitive to local minimums.

Second-order and higher order techniques (such as the Levenberg–Marquardt) can be used for updating weight, giving faster convergence at the cost of power calculation. Higher order techniques are generally more sensitive to local minimums.

3.9 Practical considerations when designing MLPs

Despite the rich theoretical literature provided for ANNs and MLPs, their use remain dependent on data and applications, their natures and the pacific environment related. This section gives some practical consideration to take into account when designing MLPs to solve function approximation or classification problems.

3.9.1 Data normalisation

First of all and before, any conceptual step, learning and validation data should be scaled in a region [−1, + 1] or [0, + 1] to ensure the proper functioning of the neuron in sensitive regions of the activation function for tangent and logarithm sigmoidal, respectively. Indeed, this remains one of the Cybenko theorem conditions; see Section 3.5. If a linear type output neuron is used, setting the scale of the output is not critical.

3.9.2 Informative training and validation data for MLPs

The heart, or the basis of any AI machine learning approach is a data base or a knowledge base, rich qualitatively and quantitatively, two key notions that will be explained later.

This remains true in the case of ANN, and MLP modelling, where data describing the process operation is essential for the phases of training and validation of the neural network. Theoretically so that an ANN, or any other ML classifier, is valid throughout the operating input/output space, training using data covering all that space must be used. In practice, this is rarely possible, and if it was, a classical IF → THEN algorithm would be sufficient to ensure all the solutions.

Luckily, in most cases the validity of models is restricted to an operating region of the input/output space in which data used for training and validation are collected.

Data is logged usually from, sensors and measures, statistical counts, observations, or by applying test protocols in the case of industrial processes, etc. These protocols consist in changing the input values, e. g., the position of a valve, a flow, electric current, etc., in order to collect output values and capture in them in the process behaviour. Regarding the models used for control and more particularly in regulation, the region is located around the operating point. Note that this is the case for most food processes, controlling a temperature, a flow, a concentration, etc.

It is often proven to be confronted with obstacles or problems that prevent the proper collection of informative data when working on industrial processes. Among others:

- Proper devices are not available, or is only available for a specified period of time in order not to stop production.
- Installation is badly and/or not properly instrumented, and does not allow harvesting of desired data. It is often the case that the management opposes any information addition of sensor and/or transmitter, which could help the test protocol,

for fear of possible breakdowns, loss of time or violation of a warranty certificate by the often very scrupulous supplier.
- Operational regions cannot be investigated for reasons of security or productivity.

These considerations are beyond the prerogatives of the modelling/control engineer, and can only be taken into account. The engineer must therefore make proof of diplomacy for negotiating a decent time in order to perform usable test protocols. In addition, he can only accommodate with existing instrumentation since the installation of new sensors or transmitters is not always possible.

For a modelling problem (function approximation), the test protocol must be long enough to allow the capture of the complete system response time. It is often the case for slow systems that the output appears as stable, where in fact it is still in the converging phase. This is why the engineer must be patient during the conduction of test protocols. Once the system response time is captured, then input variations might be performed for shorter time in order to capture the process high behaviour frequency and obtain valuable data at those frequencies and behaviours.

3.9.3 Training procedures

After choosing an appropriate topology, based on the regressor vector dimension and the complexity/order of the process or data, the MLP must undergo a training phase in order to define the network parameters, i. e., the weights and biases of each neuron, e. g., the Back Propagation (BP) learning algorithm.

After the training phase, the predictive or generalisation capabilities of the MLP must be tested on a new dataset. The procedure is called “split sample validation”.

In what follows, the problems that may be encountered during and after training, as well as using a restricted dataset, are address and solutions and are offered.

Early stopping validation

As explained in Section 3.6, the MLP parameters are estimated minimisation a quadratic criterion based on the sum squared error (SSE) and the BP algorithm. It is proven that the BP rule converge and that the SSE will keep decreasing asymptotically to zero, with the number of training iterations and epochs. The learning algorithm can, therefore, be run until there is no more perceived improvement, i. e., until a global (or local) minimum is reached.

It was, however, noted and reported in the literature that if the model is evaluated on a so-called validation set, monitoring the validation error (the error obtained from this set), that this error start by decreasing then will at a certain point deteriorate and increase. This phenomenon is termed “over-training”; see Figure 3.14 [10]. In Chi, in 1995, [11], the over-training problem based on an artificial two-input two-category

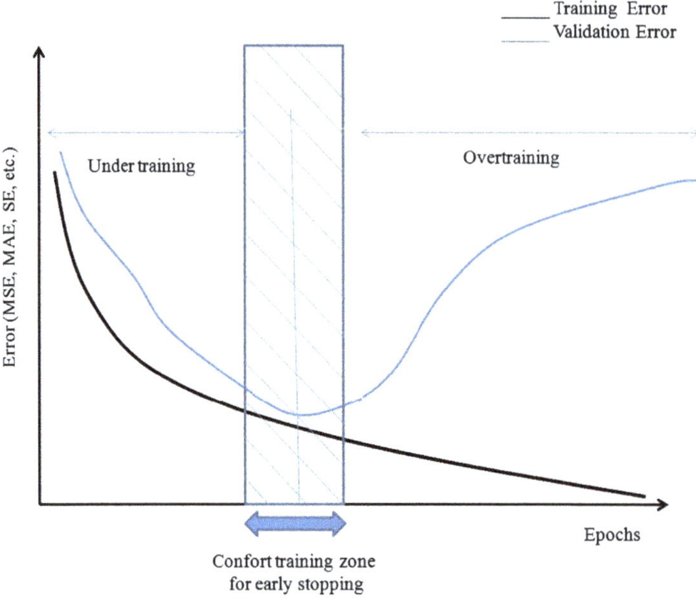

Figure 3.14: Over-training in ANNs and MLPs.

classification problem is discussed. Five solutions to the problem are presented and supported by experimental results. Sjoberg and Ljung in [4], also adress the issue of overtraining in the search for the minimum validation error.

In other words, as you learn, the MLP will tend to learn the dataset used for learning, resulting in poor generalisation and a bad prediction for a different dataset. In order to overcome this problem, a learning supervised method based on validation error is implemented. This method checks the validation error on each iteration of the learning process and compare it to its previous value in order to determine when the deterioration begins. At this point, the weight and bias values are saved. In order to make sure that this is a global minima (or at least a decent local minima), learning is not interrupted and should be performed for a sufficient number of iterations or epochs, to ensure that the validation error will no decrease again (note that this is highly improbable after long training).

The proposed solutions to avoid over-training are detailed in [11] and given in what follows:

Early stopping (ES): The most straightforward but not yet guaranteed method, as deciding when to stop training is still an issue. The stopping epoch can only be efficiently chosen using a monitoring of the validation error and repeating the training.

Pre-eliminating Confusing Patterns (PECP): The approach consists in eliminating patterns coming from different categories and confused in overlapping regions of the input/output space. The remaining training set includes then only separable patterns. Let D be the average distance between two patterns obtained from two different categories and r a product factor. For two patterns, P_iE (category i) and P_jE (category j) ($i \neq j$), let $D(P_i, P_j)$ be the distance between the two patterns, if

$$D(P_i, P_j) \leq rD_a \tag{3.50}$$

then the patterns P_i and P_j are just removed from the training set. It is frequent that some non essential categories are masked and not removed.

Contribution Factor in Training (CFTR): weighted error function might be used when a MLP classifier gets an unbalanced performance among categories or its performance is not proportional to the importance of each category, the following weighted error function is proposed:

$$E_p = \frac{1}{2}\sum_{i=1}^{m}(a_i(T_{pj} - O_{pj}))^2 \tag{3.51}$$

where a_i is some sort of importance coefficient of the ith category, that should be satisfying

$$\sum_{i=1}^{L} a_i N_i = \sum_{i=1}^{L} N_i = N \tag{3.52}$$

Penalty in testing (PIT): The commonly used testing scheme for a MLP classifier is the "winner-take-all" approach. If a classifier has been over-trained in the direction of favoring some categories, a penalty scheme might be introduced in the calculation of output activations, as follows:
Let O_i be the activation of the ith output, where p_i is the penalty factor (≤ 1.0) affected to that output. The activation of the penalised output i, denoted by O_i^p, is therefore

$$O_i^p = p_i O_i \tag{3.53}$$

Modified Training Scheme (MTRA): Modified training schemes, affects the learning rule itself.
Similar patterns produce similar activations in one or more nodes, this may lead to overtraining. The solution is then to eliminate them during training by checking the outputs for each training pattern applied to the network. The pattern is considered redundant if two or more output nodes exhibit large activation (exceeding a set threshold), and are systematically removed so that the produced error E_p does not contribute to weight change ΔW_j.
An example of the learning rule update is giving in [11].

Cross-validation

Cross-validation may bring improvements over split sample validation, in the sense that all sets of collected data are used in turn for validation. This is most effective when the size of the data space is restricted and/or limited. The method is also called Cross Model Selection.

In practice, the method divides the existing dataset D into M subsets. This results in M MLP models to be trained and validated using all M subsets; see Figure 3.15.

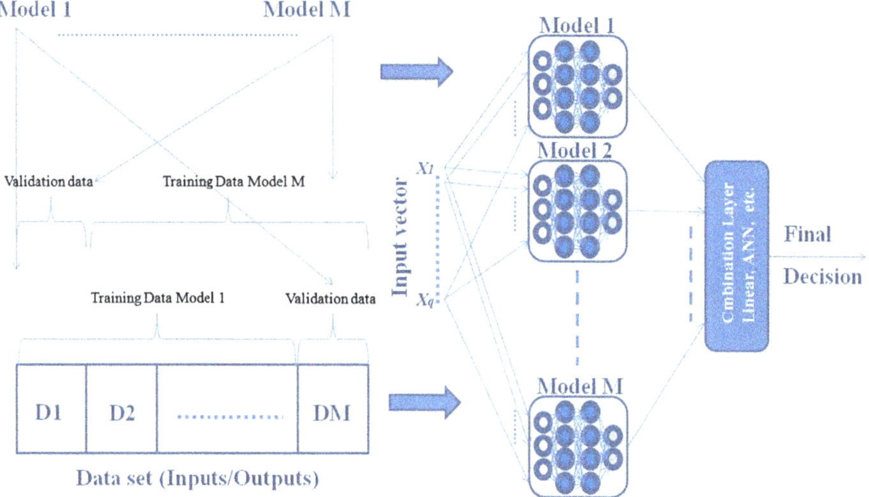

Figure 3.15: Cross-validation and multi-models.

The MLP models are provided during the training and validation procedure, each time with a new different subset selected from the dataset D. The procedure is as follows:

- The first model MLP_1, uses all subsets $D_1, D_2, \ldots, D_{M-1}$ for training while subset D_M is used for validation.
- The second model MLP_2, uses all subsets $D_1, D_2, \ldots, D_{M-2}, D_M$ for training while subset D_{M-1} is used for validation.
- The procedure is repeated until training the last model MLP_M using $D_2, D_3, \ldots, D_M$ for training and D_1 for validation.

Every time a single subset D_i is used for validation, while the rest of the subsets are concatenated and used for training. Having multiple validation estimates covering the entire data space provide a greater degree of accuracy. Note that over-training is still have to be monitored for each model.

At the end of the cross-validation M, MLP models are obtained. The designer can then choose the best or most suitable one for the operating region or input/output space area.

A second approach is to combine all the models obtained in a multi-classifier strategy, where a simple linear combination can be used for this purpose. Multi-classification and combining several ANN models (MLPs or others) might be performed using nonlinear combination approaches involving other ANNs types.

3.9.4 Topological optimisation of MLPs

Once the problems of choosing the type of network, the type of neurons as well as the number of hidden layers to be used are sliced; the remaining task of choosing an adequate topology (number of neurons in each layer) is not straightforward. There are no traces in the literature of the existence of well-defined laws or theorems giving an optimal topology for a given test protocol/data/problem. We are therefore left with the general objectives of topological optimisation, which goals are to obtain a network:

- Whose complexity is somewhere equivalent to the complexity of the problem to be solved.
- Capable of reproducing the learning results, and
- Capable of good generalisation (not too complex).

Two topological optimisation approaches for MLPs must be distinguished:

- Network growing.
- Network pruning by:
  (a) Complexity regularisation.
  (b) Sensitivity analysis.

Network growing

Using this method, the network is enlarged (an additional neuron is added) only when the goal to be reached (e. g., a specified MSE) is impossible to achieve. However, the approach has some drawbacks, and one may cite:

- Sensitivity to initial conditions.
- The position of the additional neuron cannot be specified.
- The design must have a defined objective (SSE, SQE, good interpolation, etc.).

Network punning

For this approach, the network is initially built with a significant number of neurons. Then this number is decreased by using pruning techniques. The principles of some pruning techniques are explained in what follows:

Complexity regularisation: Regularisation of complexity aims to limit the complexity of the network by a strong penalisation of weights, which do not greatly affect the output of the system. This comes to minimizing a compound cost given by

$$J_{\text{tot}} = J_e(W) + \rho J_c(W) \tag{3.54}$$

Two procedures can be used to limit network's complexity:

Weight deficiency: This method forces weights (after an initial period of convergence) to tend towards zero, unless constantly reinforced by the learning algorithm is performed. This can be formulated as follows:

$$\Delta w_{ji} = \Delta w_{ji} - \rho\, \text{sgn}(w_{ji}) \tag{3.55}$$

where ρ is a "forgetting" factor, given the following corresponding cost function:

$$J_c = \sum_{i,j} |w_{ji}| \tag{3.56}$$

This approach is sometimes termed as structural training.

Weights elimination: Here, penalty is put on complexity and is defined by

$$J_c = \sum_{i,j} \frac{1}{1 + (w_{ji}/w_0)^2} \tag{3.57}$$

w_0 is predefined, and may be interpreted as the median or mean significative weight. Variations in J_c depends here of the relative values of w_{ji} and w_0 as

$$\begin{aligned} &\text{if} \quad w_{ji} \ll w_0 \quad J_c \to 1 \\ &\text{if} \quad w_{ji} \gg w_0 \quad J_c \to 0 \end{aligned}$$

Sensitivity analysis: For this approach, a local model of the error surface is constructed, for the analytical prediction of the effects of disturbances on the network weights. The change, δJ in the cost function results from a change δW_i in the weights i, and is given by the following Taylor series:

$$\delta J = \sum_i g_i \delta w_i + \frac{1}{2} \sum_i \sum_j h_{ji} \delta w_i \delta w_j + \text{terms} \tag{3.58}$$

where g_i is the gradient vector of J, given by

$$g_i = \frac{\partial^2 J}{\partial w_i} \tag{3.59}$$

h_{ji} is the Hessian matrix of J, given by

$$h_{ji} = \frac{\partial^2 J}{\partial w_j \partial w_i} \tag{3.60}$$

in that order, two simplifications may be done:

- Suppress weight only after convergence $\Rightarrow g_i \to 0$
- Quadratic approximation $\Rightarrow$ higher order terms $\to 0$

Equation (3.58) becomes then

$$\delta J \approx \frac{1}{2}\delta w^T H \delta w \tag{3.61}$$

where H is the Hessian matrix containing all the second-order derivatives. Equation (3.61) may now be minimised subject to the following constraint:

$$\delta w_i + w_i = 0 \tag{3.62}$$

This constraint optimisation problem can be solved by using Lagrange multipliers by following the procedures described in:
- Optimal Brain Damage (OBD) introduced by LeCun et al. in 1990 [16].
- Optimal Brain Surgeon (OBS) introduced by Hassibi et al. in 1993 [17, 18].

3.10 Case study: Application of MLP for wine classification

In this section, an illustration of the usage of neural networks and more precisely a regular MLP, is given in the form of an application for wine producer's classification.

Wine is not only one of the most valuable beverages in the world with a substantial market worth around 30 billion US$, but also a cultural and traditional constant in many countries and societies. Indeed, the most important wine consumers are in Europe, with France, Italy, and Portugal having the highest per capita consumption over 35 liters per person per year, when compared with 23.9, 9.9, and 3.5 for Australia, USA, and China, respectively.

Until recently, the quality and origin of wine have always been determined by wine experts. Nowadays various advanced procedures and devices exist for the task (See E-nose and E-tongue explained and detailed in Sections 2.1.1 and 2.1.2). In this section, ANNs and most precisely the MLP is designed, implemented, and tested for the classification of wine according to three given producers, based on wine specific characteristics.

Wine may be considered as a complex product issued for a relatively long fermentation and aging process. Its composition depends on many and diverse factors such as edafoclimatic conditions, grape variety, and enological practices [21]. All of them greatly influence the wine quality, characterisation, and differentiation, and are of paramount importance for the detection of frauds [22]. The usage of those characteristics and others have been implemented in various ANN based classification and modelling approaches with all of them reported in Section 2.5.6, Table 2.6.

3.10.1 Wine dataset used

The dataset used is labelled with three classes, numbered as 1, 2, and 3 (3 types of wines belonging to three distinct producers), making the classification problem a supervised one by essence. The dataset contains 178 examples, each constituted of a characteristic vector of 13 parameters, as inputs and one output corresponding to the existing classes represented by the numbers 1, 2, and 3.

The input vector contains the proportions of the following components: alcohol, malic acid, ash, alkalinity of ash, magnesium, total phenols, flavonoids, nonflavanoid phenols, proanthocyanins, colour intensity, hue, OD280, proline. Figure 3.16 shows a sample of the dataset and the above mentioned characteristics.

Alchol	Malic_Acid	Ash	Alcalinity_of_Ash	Magnesium	Total_phenols	Falvanoids	Nonflavanoid_phenols	Proanthocyanins	Color_intensity	Hue	OD280
14.23	1.71	2.43	15.6	127	2.80	3.06	0.28	2.29	5.64	1.04	3.92
13.20	1.78	2.14	11.2	100	2.65	2.76	0.26	1.28	4.38	1.05	3.40
13.16	2.36	2.67	18.6	101	2.80	3.24	0.30	2.81	5.68	1.03	3.17
14.37	1.95	2.50	16.8	113	3.85	3.49	0.24	2.18	7.80	0.86	3.45
13.24	2.59	2.87	21.0	118	2.80	2.69	0.39	1.82	4.32	1.04	2.93

Figure 3.16: Sample of the dataset used for wine classification.

Based on those characteristics, a MLP is perfectly able, when well trained, to accurately classify any new sample presented [20]. For that purpose, the dataset is divided into training and validation sets, containing respectively 133 and 45 samples.

Before being used for training and validation of the designed MLP, the dataset needs to be standardised. Standardisation of datasets is a common, and sometimes a compulsory requirement for ANN design, as may not be trained correctly or lack generalisation if the individual features are not uniformed and look like standard normally distributed data, i. e., Gaussian with zero mean and unit variance. Indeed, if a feature is significantly different in terms of magnitude, it might dominate the discriminant function and affect the estimator generalisation on other less predominant features.

In practice, the shape of the distribution is often ignored and just transforms the data and centring them by removing the mean value of each feature, then scaling them by dividing non-constant features by their standard deviation. A standardised dataset sample is given in Figure 3.17, where the differences between initial characteristic values are reduced and homogenised, between −1 and +1, after scaling the dataset shown in Figure 3.16.

3.10.2 Creation and training a MLP for wine classification

Once the dataset is homogenised, scaled, and ready for being used to train and validate a MLP, the architecture of the latter is to be chosen. After several trials, an archi-

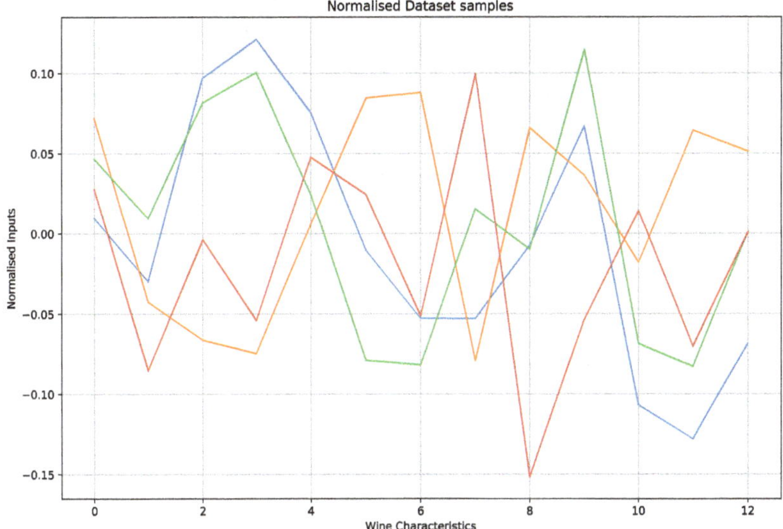

Figure 3.17: Normalised samples of the training dataset.

tecture of [13-17-20-3] is found to fit the problem complexity, where the network is composed of 13 inputs given by the dimension of the input vector, 17 neurons in the first hidden layer, 20 neurons in the second hidden layer, and 3 output neurons. The training of the MLP for 700 epochs shows the evolution of the training error displayed in Figure 3.18. The figure shows training and validation errors until the epoch 250 where

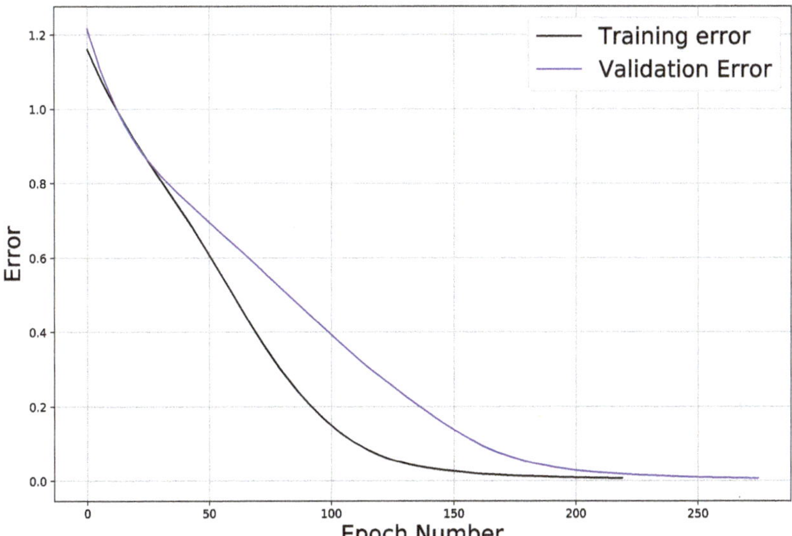

Figure 3.18: Training error given for 250 epochs.

the training process is stopped earlier than the 700 epochs initially planned. The training process is stopped as the training error of 0.001 is reached. Meanwhile, the validation error is kept at sensibly the same value without experiencing over-training. The speed of the training process is due to the limited training and validation datasets and relatively small number of training parameter (i. e., 13) for each sample.

Once the training is performed, validation using a different dataset must be performed in order to assess the generalisation capabilities of the trained MLP.

3.10.3 Validation and results assessment

Now that the model parameters are fixed after training, the MLP can be used to predict and classify the validation dataset, composed of 45 wine samples.

The procedure is simple and consist in simulating the MLP outputs when presented with the validation samples one after the other. After several trials with the same MLP topology, it is possible to reach a 100 % classification performance on the validation dataset. Indeed, the validation results shown in Figure 3.19 that all forecasted classifications represented by a blue "X", correspond and relates to all real classification represented by a black "O".

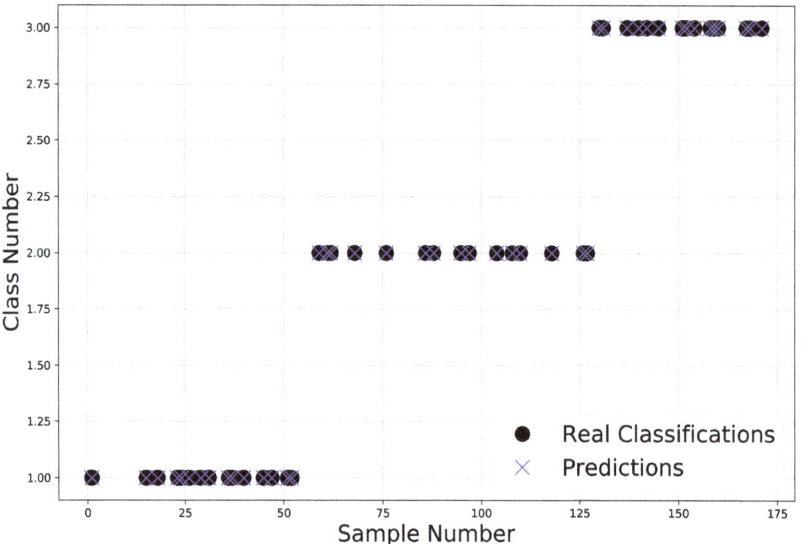

Figure 3.19: Training error given for 700 epochs.

Bibliography

[1] Ljung, L. (1999) System Identification – Theory for the User. Second edition. Englewood Cliffs, N.J: Prentice Hall.

[2] Chen, W. H., Billings, S., and Grant, P. (1990) Nonlinear system identification using neural networks. International Journal of Control, 51(6), 1191–1214.

[3] Nørgaard, M., Ravn, O., Poulsen, N. K., and Hansen, L. K. (2000) Neural Networks for Modelling and Control of Dynamic Systems, London: Springer-Verlag.

[4] Sjoberg, J. and Ljung, L. (1992) Overtraining, regularisation, and searching for minimum in neural networks, in 10th IFAC Symposium on System Identification, Copenhagen, volume 2, pp. 49–72.

[5] Box, G. E. P. and Jenkins, G. M. (1989) Time Series Analysis, Forecasting and Control. Second edition. San Fransisco: Holden day.

[6] Sjoberg, J. (1995) Nonlinear System identification using neural networks. PhD Thesis. Dept. of Electrical Engineering, Slinkoping University, Sweden.

[7] Cybenko, G. (1989) Approximation by superpositions of a sigmoidal function, Mathematics of Control, Signals and Systems 2, 303–314.

[8] Kwon, Taek Mu and Cheng, Hui (1996) Contrast enhancement for backpropagation. IEEE Transactions on Neural Networks, 7(2), 515–524. doi:10.1109/72.485685.

[9] Nawi, N. M., Hamid, N. A., Ransing, R. S., Rozaida, G., and Salleh, M. N. M. (June 2011) Enhancing back propagation neural network algorithm with adaptive gain on classification problems, Journal of Database Theory and Application, 4(2).

[10] Tetko, I. V., Livingstone, D. J., and Luik, A. I. (1995) Neural network studies. 1. Comparison of overfitting and overtraining. Journal of Chemical Information and Modeling, 35(5), 826–833. doi:10.1021/ci00027a006.

[11] Chi, Z. (1995) MLP classifiers: overtraining and solutions, in Proceedings of ICNN'95, International Conference on Neural Networks. doi:10.1109/icnn.1995.488180.

[12] Chen, T., Chen, H., and Liu, R. (1992) A constructive proof and an extension of Cybenko's approximation theorem, Computing Science and Statistics, Springer-Verlag New York, Inc.

[13] Rumelhart, W. H., Hinton, G., and McClelland, J. (1986) A general framework for parallel distributed processing. Parallel Distributed Processing, 1(2), 45–76.

[14] Plaut, D. C., Nowlan, S. J., and Hinton, G. E. (1986) Experiments on learning by back propagation.

[15] LeCun, Y., Bottou, L., Bengio, Y., and Haffner, P. (1998) Gradient-based learning applied to document recognition. Proceedings of the IEEE, 86(11), 2278–2324.

[16] LeCun, Y., Denker, J. S., and Solla, S. A. (1990) Optimal brain damage, in Advances in Neural Information Processing Systems, pp. 598–605.

[17] Hassibi, B., Stork, D. G., and Wolff, G. J. (1993) Optimal brain surgeon and general network pruning, in IEEE International Conference on Neural Networks, pp. 293–299.

[18] Hassibi, B. and Stork, D. G. (1993) Second order derivatives for network pruning: optimal brain surgeon, in Advances in Neural Information Processing Systems, pp. 164–171.

[19] Schraudolph, N. (2002) Fast curvature matrix-vector products for second order gradient descent. Neural Computation, 14(7), 1723–1738.

[20] Ozgur, C. Classification du vin avec rèseau de neurons artificiels. Gènie èlectrique et èlectronique, Universitè des sciences et technologies d'Adana, Adana.

[21] Troost, G. (1985) Tecnologìa del vino, Omega S.A., Barcelona, Spain.

[22] Pèrez-Magariñoa, S., Ortega-Herasa, M., Gonzàlez-San Josèa, M. L., and Bogerb, Z. (2004) Comparative study of artificial neural network and multivariate methods to classify Spanish DO rose wines, Talanta, 62, 983–990.

4 Radial basis function networks

So-called feedforward neural networks and neural networks with Radial Basis Functions (RBF) are two widely used parametric model classes in identification and function approximation of nonlinear systems. Indeed, these networks with a single hidden layer can approximate any continuous function having a finite number or discontinuities (Cybenko theorem [1], see Section 3.5). A clear revival of interest for RBFs was noted during the years 1990 to 2005 in various fields of application, such as signal processing, checking and diagnosing errors, approximation of time series, etc. For a given problem, the use of an RBF generally leads to a structure less complex (number of units in the hidden layer). Moreover, the computational complexity induced by their learning is less important than the one produced by the MLP network, favored by the existence of hybrid training algorithms applied to RBFs. Performances of such a network depend on the:

- Choice of the centres and widths of the radial basis functions in the hidden layer.
- Choice of the number of functions constituting the hidden layer (number of neurons) and of the estimate network settings.
- The estimating of the network parameters in the learning phase, during which a set of experimental input-output pairs are used to enable the RBF network to acquire a nonlinear input-output relationship.

4.1 Radial basis function network topology

Introduced by Powell [2] and Broomhead and Lowe [3], the RBF network (Radial Basis Function) is part of the large supervised neural networks family. It consists of three layers, Figure 4.1:

- an input layer, which retransmits the inputs without distortion to the hidden layer,
- a single hidden layer which contains RBF neurons, which are generally Gaussian and,
- an output layer of which neurons are generally driven by a linear activation function.

Each layer is completely connected to the next and there are no connections inside the same layer [5]. The network shown in Figure 4.1, consists of N input neurons, M hidden neurons, and J output neurons. The output of the mth neuron from the hidden layer is given by

$$y_m^q = \left[-\frac{\|x^q - v_m\|}{2\sigma_m^2} \right] \tag{4.1}$$

v_m is the centre of the mth neuron of the hidden layer or of the mth Gaussian neuron and σ_m is the width of the mth Gaussian function. The output of the output layer, j^{th}

https://doi.org/10.1515/9783110646054-004

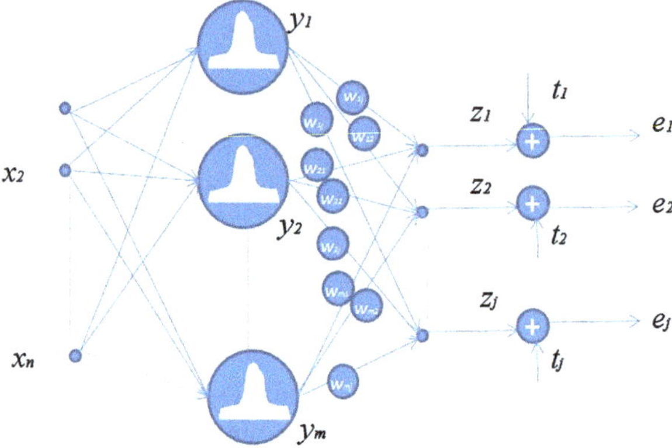

Figure 4.1: Radial basis function architecture.

neuron is given by

$$Z_j^q = \frac{1}{m}\left[\sum_{1}^{M} w_{mj} y_m\right] \tag{4.2}$$

w_{mj} are the weights connecting the hidden layer to the output neurons. An example of the value of σ for a given Gaussian function is given by Figure 4.2.

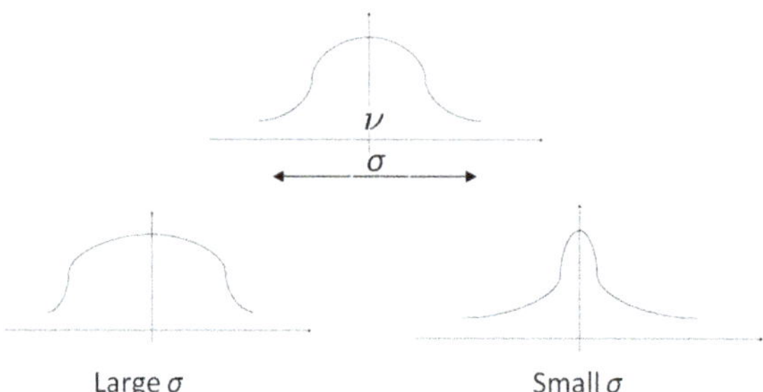

Figure 4.2: Radial basis function shapes.

Another difference between MLPs and RBF networks is the nature of the separation introduced by the RBF being of local type by comparison to the global aspect of the separation introduced by a MLP. Figure 4.3 explains the two types of separations.

Indeed, on one hand and from Figure 4.3(a) the linear separation produced by a single threshold or sigmoid neuron implemented in a MLP type of network is clearly seen for a 2-dimensional problem. Separation using a line is infinite by nature and could implement a global classifier if the problem is linearly separable (see Section 1.7) and more than one line (or neurons) if the problem is nonlinearly separable. On the other hand, Figure 4.3(b) shows the local separation implemented by a circle given

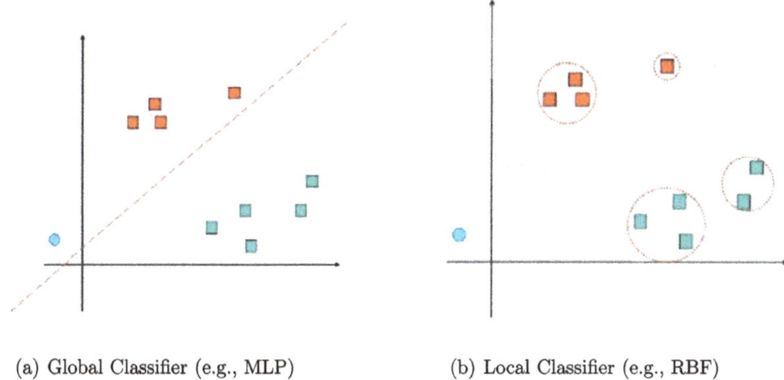

(a) Global Classifier (e.g., MLP) (b) Local Classifier (e.g., RBF)

Figure 4.3: Difference between MLP and RBF classification type.

by a radial or gaussian function in a 2-dimensional problem. Indeed, a single neuron implementing a Gaussian function includes a number of similar data elements or a complete cluster if this one is isolated from the rest.

4.2 Radial basis function network's training algorithm

RBF networks learning algorithm was first presented by Moody and Darken [6], and consists of adjusting four main parameters:

- The number of elements (neurons or Gaussians) on the hidden layer.
- The position of the centres of these Gaussians.
- The dimension (Width and shape) of those Gaussians.
- The connection weights of the output layers (linear neurons).

The goal always consisting in minimizing the Sum Squared Error (SSE), or another type of distance, is calculated between the outputs obtained from the network and those desired:

$$\text{SSE} = \sum_{i=1}^{Q}\sum_{j=1}^{J}(t_j^q - z_j^q)^2 \tag{4.3}$$

For the RBF network, the adjustment of the weights w_{mj} connecting the hidden layer neurons to that of the output is achieved by the Widrow–Hoff rule, and is performed as follows:

$$w_{mj}^{i+1} = w_{mj}^{i} + \mu(t_j - z_j)y_m \tag{4.4}$$

t_j is the desired output of the jth neuron, z_j is the calculated output of the jth neuron, y_m is the output of the mth neuron of the hidden layer and μ is the learning step whose value is between 0 and 1. The parameters to be set (for Gaussian RBFs) are:

- The Gaussian function centres v_i;
- The Gaussian function widths or dispersion σ_i, and
- The weights w_{mj}.

The main advantage of RBF is that it is possible to simplify the learning process by dividing the work in three distinct phases:
- positioning of centres,
- determining the width of the Gaussian nuclei, and
- weight adjustment.

There are several learning strategies for training RBF networks [7]. Depending on how the centres are specified, Ghosh and Nag [8]cite three:

Randomly selected centres: choose centres and dispersions following analysis of the data, the weights are the solution of the linear equations using equation (4.4).

Centres obtained by unsupervised learning: estimation of the Gaussian's centres by categorization using Kmeans. Dispersions can be chosen as the average distance between centres neighbors, and the weights are the solution of the linear equation (4.4).

Centres obtained by supervised learning: All of the parameters are adjusted to minimize an error function. The problem comes then to the global optimisation problem adjusting all the RBF network parameters.

In classification tasks, the number and the centres of the Gaussians are chosen usually using Kmeans-type data gathering techniques. The widths of the Gaussians are calculated by the average value of the distances separating all the examples at the corresponding centre. K-means are iterative methods for separating a series of vectors into different clusters and each cluster is represented by a centre.

The use of RBF networks requires, in summary, two major decisions: What should be the transfer function? and if it is a Gaussian, how to choose the settings. There are no definite answers to these questions and the solutions remain empirical. The width of the Gaussian should generally be wider than the distance between two points in space but smaller than the width of the cluster diameter.

How much centre should I use and where? Generally, the least possible centres are used, of course, trying to keep as much precision as possible. The zero error target can be easily approached for training data, but validation must be done on a different dataset, as for back-propagation learning seen previously, Section 3.9.3.

4.2.1 K-Means algorithm

The K-means algorithm, mainly use as a first step for training RBF networks and finding cluster centres and width in order to fix neurons centres and widths is as follows:

- Randomly choose the centre of each of the K clusters.
- Assign each object to the cluster whose centre is closest to it.
- Recalculate the positions of the new centres.
- Recalculate the positions of the new centres. Repeat steps 2 and 3 until convergence, i. e., until a number of iterations maximum is reached.

The K-means algorithm has advantages such as: rapid convergence, ease of implementation, and the ability to process relatively large databases, although this has to be reconsidered with the new dimension of datasets and data warehouses. Its disadvantage lies in the need to know a priori the number of clusters and remains a linear approach, although the optimal number of clusters may be identified using a number of indexes such as: silhouette, elbow, etc. An alternative to K-means may be a nonlinear approach that will determine automatically the number of clusters. AI and neural approaches might be considered as clustering solutions, such as the one given by Kohonen maps (see Chapter 5).

4.2.2 Radial basis function curse of dimensionality

A modern day classification problem often consists of possibly hundreds to thousands of variables, with a limited number of samples due to the workload of data acquisition.

Such high-dimensional data with small sample sizes makes pattern classification methods suffer from the "curse of dimensionality". This is even highlighted in the case of RBF networks and Gaussian functions, as the concentration of the norm phenomenon results in the fact that Euclidean norms and Gaussian kernels, suffers greatly in high-dimensional spaces.

Many classification algorithms suffer when the number of features in the data is sufficiently large. This problem is further exacerbated by the fact that many features during the learning task may either be irrelevant or redundant to others with respect to predicting the class of an instance [4].

Data dimensional reduction is, therefore, a good solution for high-dimensional data processing. As one of dimensionality reduction methods, feature extraction methods have been explored, for example, Principal Component Analysis (PCA) [9], High-order Statistics (HOS) [10], Independent Component Analysis (ICA) [11], etc.

4.3 Classification of simple examples using RBF: the XOR problem

As in Section 3.7, where a single layer MLP with two neurons in the hidden layer is used to classify ones and zeros in a two-dimensional XOR function, the problem is revisited here using an RBF network as a classifier. The same configuration is required, for instance, an RBF containing two neurons in the hidden layer and a combination linear neuron as an output.

Let us recall the two-dimension XOR problem, with its are four patterns namely: (00) (01) (10) (11) given as outputs the following vector: $\hat{y}^T = [0110]$. The RBF model should then construct the output vector a pattern classifier that produces the output $\hat{y}$ with 0 for the input patterns (0,0), (1,1), and the output 1 for the input patterns (0,1), (1,0).

To perform the XOR classification using an RBF network, and as stated in Section 4.2 one must begin by choosing how many basis functions are needed. Given there are four training patterns and two classes, $M = 2$ is the simplest network that is able to solve the XOR problem. Note that this is similar to the solution given by a two neurons in a single layered MLP, Section 3.7. The basis function centres and widths need to be determined either by an intuitive choice in the case of simple well-visualised problems, or using K-means if the data are of higher dimension and nonlinearly scattered. The two separated zero targets seem a good "random" choice, so $c_1 = (0, 0)$ and $c_2 = (1, 1)$ and the distance between them is dmax $= v2$, giving the following basis functions, shown graphically in Figure 4.4(a):

$$\phi_1(x) = e^{-\|x-c_1\|} \tag{4.5}$$

and

$$\phi_2(x) = e^{-\|x-c_2\|} \tag{4.6}$$

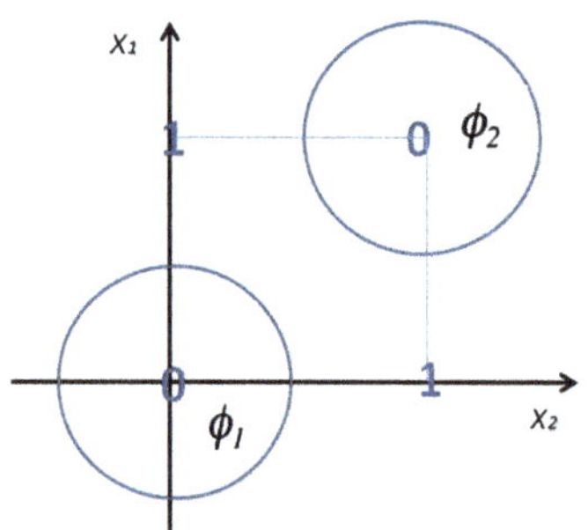

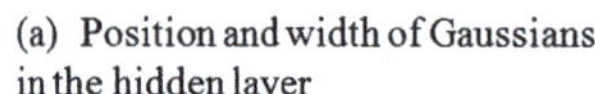
(a) Position and width of Gaussians in the hidden layer

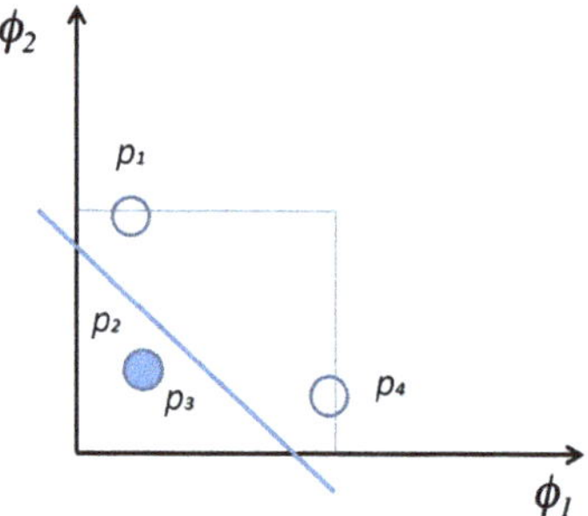

(b) Outputs of the hidden layer and output classification

Figure 4.4: The XOR problem solved using a two neurons RBF network.

The outputs of functions ϕ_1 and ϕ_2 are given in Table 4.1.

The hidden layer with the two Gaussian functions will be hopefully sufficient to transform the problem a nonlinearly separable problem into a linearly separable one,

Table 4.1: Outputs of the RBF network hidden layer.

Pattern	x_1	x_2	ϕ_1	ϕ_2
1	0	0	1.0000	0.1353
2	0	1	0.3678	0.3678
3	1	0	0.3678	0.3678
4	1	1	0.1353	1.0000

able to be treated by the single output combination neuron. Since the input/output space is only two-dimensional, it is easy to plot the activations in order to highlight how the existing four have been transformed 4.4(b). It is clear that the patterns are seen as linearly separable in the output layer, and in this case, a single neuron is well able to separate the two classes.

In the current example, there is just one output $y(x)$, with two weights w_j, one from each hidden unit j, and one bias $-b$. Thus, the network's input-output relation for each input pattern x is given by

$$y(x) = w_1\phi_1(x) + w_2\phi_2(x) - b \tag{4.7}$$

Thus, in order to make the outputs $y(x^p)$ match the targets $\hat{y}^p$, four equations are to be satisfied:

$$\begin{aligned}
1.0000 \ w_1 + 0.1353 \ w_2 - 1.0000 \ b &= 0 \\
0.3678 \ w_1 + 0.3678 \ w_2 - 1.0000 \ b &= 1 \\
0.3678 \ w_1 + 0.3678 \ w_2 - 1.0000 \ b &= 1 \\
0.1353 \ w_1 + 1.0000 \ w_2 - 1.0000 \ b &= 0
\end{aligned}$$

The solution of the above equations give $w_1 = w_2 = -2.5018$ and $b = -2.8404$.

4.4 Case study: classification of oil origin using spectral information and RBFs

Similar to the application example, given in Section 3.10 for the classification of wine producers regarding wine characteristics; here, classification of olive oil country of origin regarding their characteristics obtained from spectral analysis is presented. The classifier used this time is a RBF network, which takes as inputs the result of the molecular Fourier transform infrared spectroscopy technique (FTIR) that has been shown to perform well for the analysis of olive oils. It has been used to detect adulteration with other vegetable oils [13] and to quantify free fatty acids [14]. Moreover, it was proven that FTIR spectroscopy and multivariate analysis are well able to distinguish the geographic origin of extra virgin olive oils, as shown in [12].

In this section, the work of [12] is revisited in order to prove the applicability and the strength of RBF networks as a nonlinear classifier instead of multivariate analysis for detecting the country of origin of olive oil sample.

The major advantage of FTIR over other chromatographic approaches is that no complex sample preparation, in terms of extraction, separation, etc. is required. The FTIR analysis used to constitute the database used in this section is relatively straightforward with the complete protocol taking around 20 minutes per sample [12]. Provided it has the required discriminatory power, FTIR potentially offers enough data per sample, permitting to a classifier to correctly determine and verify the country of origin.

4.4.1 Database and available data

Using FTIR, it was possible to obtain 60 authenticated samples of virgin olive oils. The samples are originating from four European producing countries, and obtained from the International Olive Oil Council in Madrid. Table 4.2 details the dataset in terms of numbers of samples from each country acquired in two discrete periods of around 2 weeks duration. The data nature consists of an absorbance spectrum, collected from each sample during each period of time.

Table 4.2: Numbers of extra virgin olive oils from each country of each origin.

Group designation	Country of origin	Number of samples
1	Greece	10
2	Italy	17
3	Portugal	8
4	Spain	25
Total		60

The concatenation of the obtained data from both acquisition periods gave a data set of 120 absorbance spectra in total (i. e., two acquisitions from each of the 60 existing samples).

Prior and between two spectral acquisitions, samples were stored in the dark at ambient temperature and the spectrometer was stored at 21 °C in order not to alter the temperature at which the spectral acquisition is performed and samples were allowed to settle to this temperature immediately before analysis. More details on how the samples where obtained are giving in [12].

All 60 spectra are shown in Figure 4.5. It can be seen that spectral quality is high with reduced or nonexisting noise effect. When it comes to differentiating between

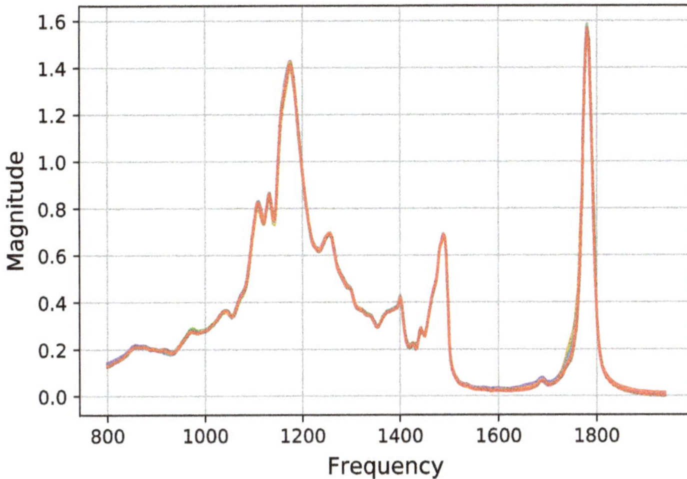

Figure 4.5: Raw spectral olive oil data for 60 samples.

sample types, it is clear that little can be gained from examining the raw spectral data visually or calculating the distance between different spectra samples.

It is clearly seen from Figure 4.5 that olive oil spectra characteristics appear virtually indistinguishable from one another, at least visually.

The spectra are usually dominated by absorptions arising from triglycerides, which constitute the major component of olive oils. In the presence of these strong absorptions, it is very difficult to see more subtle spectral contributions arising from, for example, differences in the fatty acid composition or from the nonglyceridic minor components. For that reason, [12] apply a number of multivariate analysis techniques in order to help separate spectra samples based on subtraction of the mean value and data centring. The multivariate model presented then investigates if the information present in these features are informative enough to distinguish systematically between the groups of similar clusters. The challenge for the RBF network is to distinguish between oil sample spectra analysis using the raw, visually indistinguishable data.

The raw data set used for training and validation of the RBF network classifier consists of a data matrix of 570×60. Where each of the 60 spectral data samples is constituted of 570 spectral contribution parameters, measured in the frequency range 800 to 2000. Figure 4.6 shows a part of the data set in numbers for 5 oil sample obtained in "Python". The magnitude spectral magnitude ranges from 0 to 1.6 scattered across the spectral frequency range: 800–2000 Hz.

	Country	p1	p2	p3	p4	p5	p6	p7	p8	p9	...	p561	p562	p563	p564	p565
0	1	0.130412	0.130675	0.132017	0.133824	0.136095	0.138944	0.141723	0.144136	0.146431	...	0.007770	0.007493	0.007512	0.007548	0.007391
1	1	0.128602	0.128790	0.130022	0.132012	0.134427	0.137070	0.139646	0.142338	0.144923	...	0.011095	0.010938	0.010817	0.010594	0.010379
2	1	0.126175	0.126733	0.128244	0.129893	0.131755	0.134528	0.137512	0.140429	0.142964	...	0.004055	0.004013	0.003950	0.003777	0.003405
3	1	0.127060	0.127551	0.128900	0.130609	0.132956	0.135759	0.138407	0.141162	0.143977	...	0.006990	0.006840	0.006725	0.006725	0.006642
4	1	0.127010	0.127513	0.128898	0.130888	0.133359	0.135990	0.138718	0.141501	0.143928	...	0.003651	0.003601	0.003500	0.003299	0.003043

5 rows × 571 columns

Figure 4.6: Part of the dataset used for training the RBF network.

4.4.2 Training and validation of the RBF network using random centring

Prior to training, and as explained in any ANN classification approach, the data are divided into a training and a validation set. In this case, data separation is performed randomly in order to maximize the network generalisation capabilities. The training set obtained contains 44 samples, where the rest 16 samples are used for validation.

The RBF modelling approach taken here is the simplest one depicted in Section 4.2. Here, the radial basis function centres are randomly chosen after defining their number (the size of the hidden layer). The weights of the output linear neuron are the solution of the linear equations using the update rule, equation (4.4). Note that in the classification problem stated, the output is rounded to the nearest value in order to fit one of the four output classes. The overall learning approach is the simplest one, and does not involve any unsupervised learning of Kmeans type, and will further test the capabilities of RBF networks to solve the classification problem based solemnly on raw spectral data.

In order to find a suitable architecture (in this case the number of neuron in the hidden layer), a network growing approach is adopted. Starting from a small number of neurons in the hidden layer, the performance of the RBF network is assessed on the basis of the validation error rate MAPE (Mean Absolute Percentage Error), and the operation is repeated every time with a larger number of neurons. Training is performed here in python using one of the RBF libraries available in [15, 16].

Table 4.3 shows a number of architectures results. A crucial point has to be noted for the random approach is that the results are very much sensitive to initial conditions that remain randomized. Therefore, the results can be significantly different for the same architecture when ran twice.

From Table 4.3, it can be clearly seen that the training MAPE reaches 0 % with a hidden layer containing 80 neurons, while validation MAPE reaches an optimal value of 6,25 % MAPE (one misclassification for the 16 validation outputs) a hidden layer composed of 140 neurons. The results of the optimal architecture are shown in Figure 4.7, where only one sample from class 4 is misclassified as class 3. It is fortuitous to test larger network as the performances will start decreasing due to over-fitting.

Note that due to the double random nature of data selection between the training and validation datasets and distribution of the Gaussian functions through the

Table 4.3: Performance of different RBF networks using network growing.

Network	Size of the hidden layer	Training MAPE	Validation MAPE
1	20	22,72	45,7
2	40	2.27	37,5
3	60	2.4	31,25
4	80	0	37,5
5	100	0	25
6	120	0	12,5
7	140	0	6,25
8	160	0	12,5
9	180	0	12,5

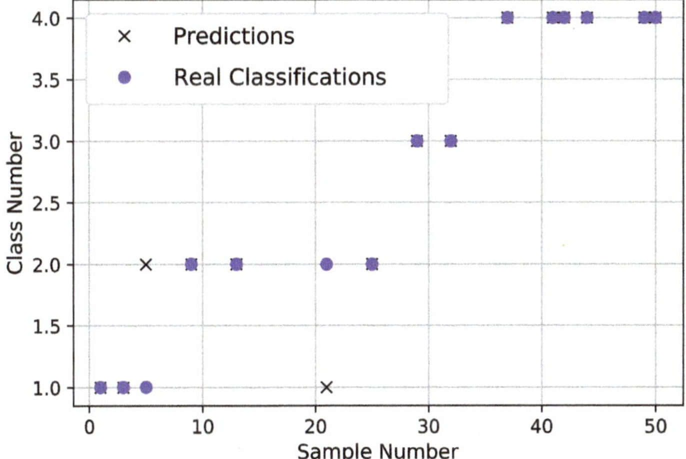

Figure 4.7: Best results obtained for a RBF 570-140-1 architecture.

input/output space, it is close to impossible to obtain the same training and validation results given for two different runs of the programme. One random dimension may be lifted in the case of adopting an unsupervised clustering approach prior to classification, using for example K-means or Kohonen maps to determine the cluster centres and tough the Gaussians coordinates. The selection of the validation and training sets should however remains random in order to ensure fairness across the existing classes. It is not impossible to obtain a 100 % validation rate when repeating the simulations over and over. However, the corresponding RBF network will not ensure zero error if new sensibly different samples are presented.

The results obtained by an "equivalent", in size terms, MLP network are "equivalent" and close in terms of results to the one obtained here by an RBF network. Indeed, a MLP of 570-140-1 topology gives the validation error of 6.25 % representing one

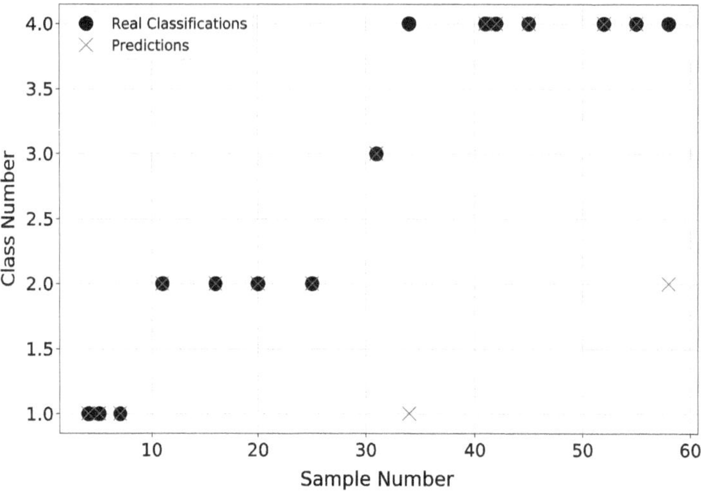

Figure 4.8: Best results obtained for a MLP 570-140-1 architecture.

misclassifications in class 4, over the 16 validation samples; see Figure 4.8. The ANN approach offers a classification solution close to a 100 % rate obtained on raw data with no extra pretreatment nor any multivariate analysis techniques.

Bibliography

[1] Chen, T., Chen, H., and Liu, R. (1992) A constructive proof and an extension of Cybenko's approximation theorem, in Computing Science and Statistics, pp. 163–168.

[2] Powell, M. J. D. (1985) Radial basis functions for multivariable interpolation A review, in Algorithms for Approximation, New York: Oxford University Press, pp. 143–167.

[3] Broomhead, D. S. and Lowe, D. (1988) Multivariable functional interpolation and adaptive networks. Complex Systems, 2(3), 321–355.

[4] Carse, B. and Fogarty, T. C. (1996) Tackling the "curse of dimensionality" of radial basis functional neural networks using a genetic algorithm, in Parallel Problem Solving from Nature — PPSN IV, pp. 707–719.

[5] Park, J. and Sandberg, I. W. (1993) Approximation and radial-basis-function networks. Neural Computation, 5(2), 305–316.

[6] Moody, J. and Darken, C. J. (1989) Fast learning in networks of locally-tuned processing units. Neural Computation, 1(2), 281–294.

[7] Howlett, R. J. and Jain, L. C. (2013) Radial Basis Function Networks 2: New Advances in Design, vol. 67. Physica.

[8] Ghosh, J. and Nag, A. (2001) An overview of radial basis function networks. In: Howlett R. J., Jain L. C. (eds.). Radial Basis Function Networks 2. Studies in Fuzziness and Soft Computing, vol. 67. Physica, Heidelberg.

[9] Wold, S., Esbensen, K., and Geladi, P. (1987) Principal component analysis. Chemometrics and Intelligent Laboratory Systems, 2(1-3), 37–52.

[10] Porat, B. and Friedlander, B. (1990) Direction finding algorithms based on high-order statistics, in International Conference on Acoustics, Speech, and Signal Processing, pp. 2675–2678.
[11] Comon, P. (1994) Independent component analysis, A new concept?, Signal Processing, 36(3), 287–314.
[12] Tapp, H. S., Defernez, M., and Kemsley, E. K. (2003) FTIR spectroscopy and multivariate analysis can distinguish the geographic origin of extra virgin olive oils. Journal of Agricultural and Food Chemistry, 51(21), 6110–6115.
[13] Lai, Y. W., Kemsley, E. K., and Wilson, R. H. (1995) Quantitative-analysis of potential adulterants of extra virgin olive oil using infrared spectroscopy. Food Chemistry, 53, 95–98.
[14] Bertran, E., Blanco, M., Coello, J., Iturriaga, H., Maspoch, S., Montoliu, I. (1999) Determination of olive oil free fatty acid by Fourier transform infrared spectroscopy. Journal of the American Oil Chemists' Society 76, 611–616.
[15] Arriaga, O. (2018) RBF-Network Minimal implementation of a radial basis function network. Available at https://github.com/oarriaga_Description.
[16] Vidnerová, P. (2019) RBF-Keras: an RBF Layer for Keras Library. Available at https://github.com/PetraVidnerova/rbf_keras.

5 Self-organising feature maps or Kohonen maps

Self-organizing maps commonly designated by SOFM (for Self-organizing Feature Maps, or SOM for Self-organizing Maps) were introduced by T. Kohonen in 1981 [2], inspired by how human brain topographic maps work, where near points in the human body are represented by groups of nearby neurons in the brain. These maps are not uniform, and are the most sensitive surface of the human body and are represented by a cluster containing the largest number of neurons. From an IT point of view, we can translate this property as follows: suppose we have a set of data that we want to classify. We are looking for a mode of representation such as objects neighbours are classified in the same class or in neighbouring classes. This type of network of artificial neurons has largely shown its effectiveness in the classification of multi-dimensional data, but unfortunately it has been ignored for many years despite his great interest after the perceptron limitations presented by Minsky and Papert. The principle of Kohonen maps is to project a complex data set on a reduced dimensional space (2 or 3). This projection allows to extract a set of vectors called referents or prototypes; see Figure 5.1. These prototypes are characterised by simple geometric relationships. The projection of data by a SOM is performed while maintaining the most important topology and metrics of the data input when displayed, i. e., close data (in the input space) will have close representations in the output space and will therefore be classified in the same cluster or in neighbouring clusters [1–4].

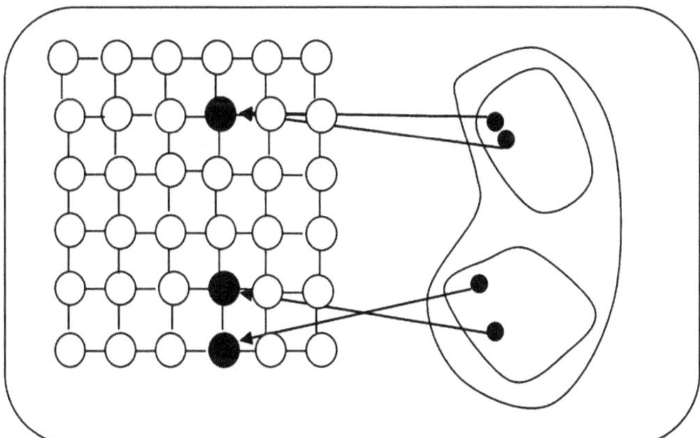

Figure 5.1: Neighbouring neurons on the map represent fairly "close" objects in the input space.

https://doi.org/10.1515/9783110646054-005

5.1 Self-organising maps architectures

The Self Organising Map (SOM) is a grid (or network) of small dimension containing a number of M neurons. In most applications, only two or at most three, dimensional grids are considered, since grids larger than 3 are difficult to visualise [5]. If the visualisation is not necessary for a given application, grids with size greater than three can be used [6]. Neurons are usually arranged either in a hexagonal or in a rectangular shape (see Figure 5.2), other topologies remains however, possible, but will not be discussed in this chapter. A Kohonen's map is usually composed of two layers of neurons: an input layer and an output layer. In the input layer, each object to be classified is represented by a multi-dimensional vector. The topological layer or the output layer is composed of a lattice of neurons according to a given geometry [7, 8]. Each neuron in the topological layer is totally connected to the neurons of the input layer $w_{.i} = (w_{1i}, \ldots, w_{ni})$. Weight vectors of these connections form the referent or prototype vector associated with each neuron, which is the same size as the input vectors. The dimension of input vectors (called "entry dimension") is generally much larger than that of the grid (called "output dimension"). Consequently, the SOM is called a "vector projection" algorithm, because it reduces the dimension of the input space (more than 2 dimensions) to the dimension of the grid (usually of dimension 2 or 3).

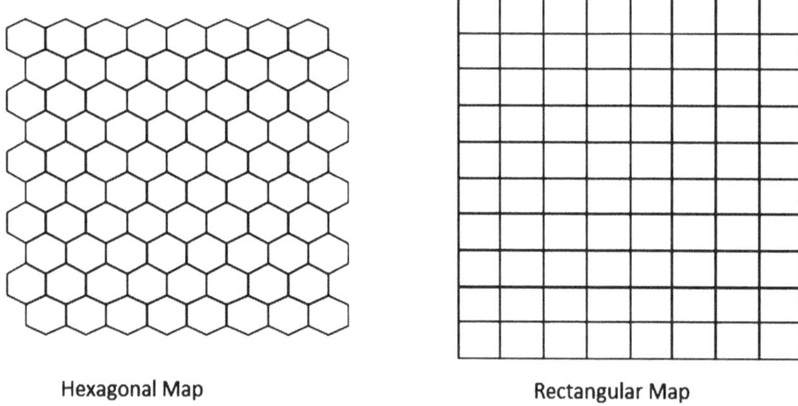

Figure 5.2: The most used topological forms of Kohonen maps: (a) rectangular (b) hexagonal.

The use of the neighbourhood notion introduces topological constraints into the final geometry of the Kohonen maps. The hexagonal architecture is the basis of most applications where, rectangular grids are also used, but their topology differs from traditional networks (Figure 5.2). The position of neurons in the network, especially the distances between them and the neighbourhood relationships are very important features and parameters for the learning algorithm. Therefore, the architecture of a

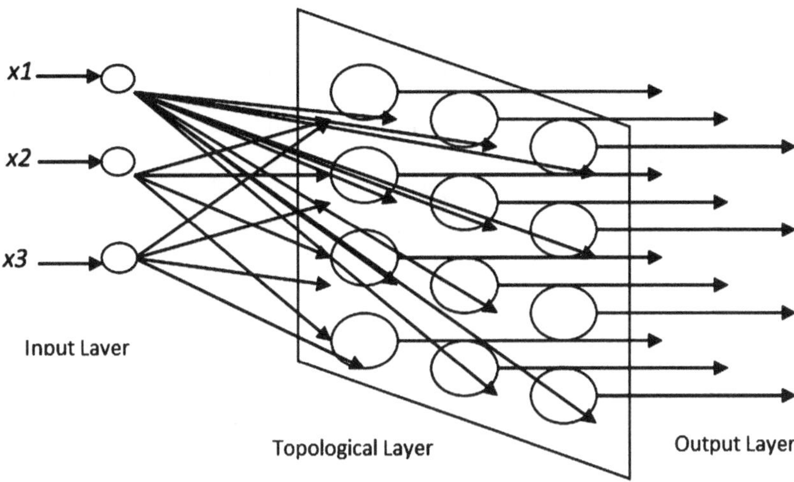

Figure 5.3: SOM structure.

two-dimensional rectangular grid Kohonen map is shown in Figure 5.3. This architecture is composed of an input layer of dimension $M = 3$ and a topological layer of dimension $L = 4\times3 = 12$ neurons. An entry vector $x(t) = [x_1, \ldots, x_M]^T$ is projected to the output layer. Each SOM input is connected to all neurons by corresponding weights, w_{ji} where $j = 1, \ldots, L$ and $i = 1, \ldots, M$. Thus to each neuron of the SOM a weight vector of dimension M is assigned $w_j = [w_{j1}, \ldots, w_{jM}]^T$.

5.2 Self-organising maps training

Once the referring vectors or prototypes are initialised, learning may be performed. The SOM is very robust when it comes to initialisation parameters, at least more than MLP and RBF networks, but good initialisation allows the learning algorithm to converge faster to a good solution. The set of learning individuals is presented to the SOM algorithm. This process is repeated for t learning steps. A complete learning cycle (when all samples have been submitted) is called an "epoch". The number of learning iterations t is an integer multiple of the number of epochs. Each iteration may be divided into two stages: a competition stage between the neurons which determines the region of the grid to be adjusted, and an adaptation stage, where the neurons weights of the selected zone are updated.

The principle of the two learning phases is illustrated in Figure 5.4. The SOM algorithm may be considered as a nonlinear variant of the K-means algorithm, in the sense where for an iteration t, the SOM learning algorithm does not only modify the centre selected as the closest to an input individual, but also the neighbouring centres for a fixed neighbourhood graph. During the learning phase, the self-organisation

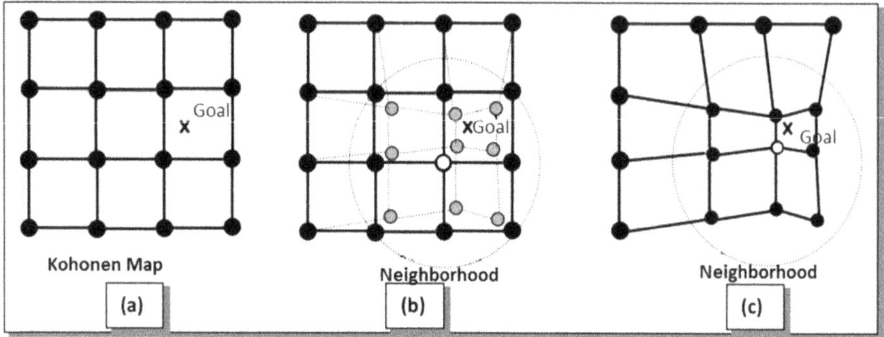

Figure 5.4: Learning steps during SOM training: (a) Initial state, (b) state at step k, (c) state at step $k+1$.

process makes it possible to focus the connection weights adaptation, mainly on the most "active" map region. This region of activity is chosen as being the neighbourhood associated to the neuron whose state is the most active, speaking therefore of a winning neuron. The criterion for selecting the winning neuron is to search for the one with the closest weight's vector from the individual presented, in terms of Euclidean distance [9, 1].

Algorithm: Pseudo-code of the SOM learning algorithm. Learning ends when the number N of epochs is reached [1].

1. For all epochs do:
2. For all inputs do:
3. Compute all the distances between the input vector and all the Map weights
4. Competition: selection of the winning neuron
5. Cooperation: adaptation of the wining neuron's weight to its neighbours,table
6. Adaptation of learning parameters
7. end for
8. end for

5.2.1 Competition phase

The competition phase between neurons is based on a discriminating function, usually the Euclidian distance between the input and weight vector. The winner of the competition is the neuron which has the greatest value returned by the discriminant function. If we take n-dimensional space for the input data set, a vector input X randomly selected from input data such that $X = [x_1, x_2, \ldots, x_n]^T$ where T denotes the transposed matrix. The synaptic weight vector of each neuron has the same dimension as the input vectors will be represented by $W_j = [w_{j1}, w_{j2}, \ldots, w_{jn}]$, where $j = 1.2, \ldots, m$, where the total number of neurons in the network is represented by m. At an instant

t, a vector $X(t)$ taken from the input space distribution is selected, and all of the neurons in the grid are then put in competition. This competition is set looking for the winning neuron, i. e., closest to the input vector. In other words, among all neurons on the map, the winning neuron noted c, is the one whose distance between its vector synaptic weights and the input vector is the lowest. This neuron is labelled "winning neuron" and often noted by BMU (Best Matching Unit). Formally, the BMU is defined as the neuron that checks the following equation:

$$|X(t) - W_j(t)| = \min_{j \in m} \|X(t) - W_j(t)\| \tag{5.1}$$

where $\|.\|$ is the distance measure.

The winning neuron for an input vector is also called the map's excitation centre. The distance generally used between the vectors X and W is the Euclidean distance, but any other type of distance can be used.

5.2.2 Adaptation phase

In order for similar input vectors to be mapped to the same neuron or to nearby neurons on the map, not only the winning neuron weights, but also its neighbours need to be updated. This action aims to force the weights of neighbouring neurons to get closer to the input vector therefore providing more similar weight vectors, and thus, similar inputs are mapped to neurons nearby on the map (Figure 5.4). The synaptic weight vectors W_j of the neuron with the index j and its neighbours on the self-organising map are updated by error correction (the error is defined as the distance between the vector x and the reference vector W_j of the considered neuron):

In expressions (5.2) and (5.3), $a(t)$ represents the learning coefficient which decreases over time to allow better adjustment of the weights. $h_{cj}(r(t))$, is the core of the neighbourhood around the winning neuron c, with a neighbourhood radius $r(t)$. The adaptation of each neuron weights is carried out according to the position of the neuron in the grid by comparing to the winning neuron.

During learning, the neighbourhood size of the BMU, which determines the active area, decreases over time. The time course of the learning coefficient is illustrated on Figure 5.5. Thus, the learning parameters are gradually modified, starting with a coarser initial phase with a large area of influence and rapid evolution of prototype vectors until reaching a fine update phase with a small neighbourhood radius and prototype vectors which adapt slowly to samples:

$$W_j(t+1) = W_j(t) + \Delta W_j(t) = W_j(t) + a(t)h_{cj}r(t)[X(t) - W_j(t)] \tag{5.2}$$

with

$$\Delta W_j(t) = a(t)h_{cj}r(t)[X(t) - W_j(t)] \tag{5.3}$$

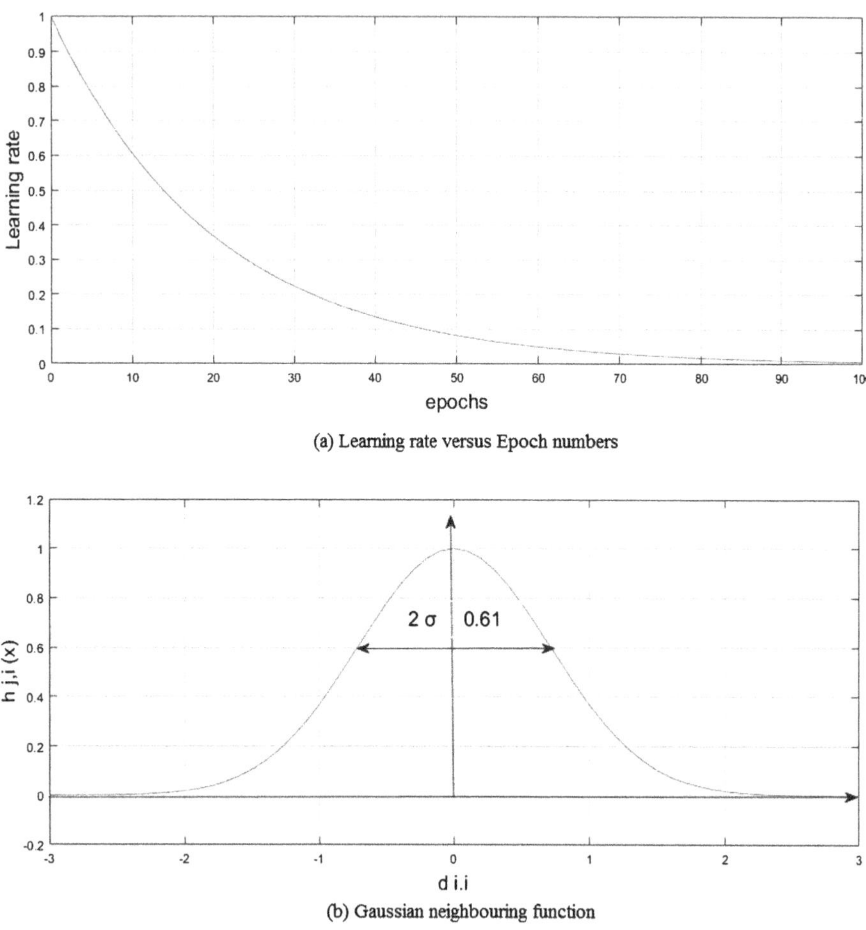

(a) Learning rate versus Epoch numbers

(b) Gaussian neighbouring function

Figure 5.5: Evolution of the parameters of a Kohonen map during learning. (a) the evolution of learning coefficient during learning, (b) The shape of the neighbourhood function for a radius of s = 0.61.

The neighbourhood function generally used in most applications is the Gaussian function. This function is centred on the neuron declared winner after the competition phase, which followed the presentation of an entry vector. The role of the modification applied to the neighbourhood chosen means bringing the selected weight vectors closer to the example presented. Therefore, the neuron whose weight vector is close to the input vector is updated so that it becomes even closer. The result is that the winning neuron is more likely to win the competition another time if a similar input vector is presented, and less likely if the input vector is completely different from the previous vector.

As previously stated, the neighbourhood function takes into account the distance between the neighbourhood neurons and the winning one to modulate the correction of neuron synaptic weights i at the instant or sample t. Let c_i be the distance between the winning neuron of index c and a neighbouring neuron of index i. This distance is not calculated in the entry space but in the topological map space:

$$\delta_{ci}^2 = \|c - i\|^2 \tag{5.4}$$

The neighbouring function $h_{ci}(t) = e^{\frac{-\delta_{ci}^2}{2r(t)^2}}$, where $r(t)$ is the neighbouring radius. This radius can be expressed by the following expression:

$$\delta_k = \delta_{ki}\left(\frac{\delta_f}{\delta_i}\right)^{\frac{k}{k_{\max}}} \tag{5.5}$$

The learning process is interrupted if one of the following conditions is met:

- the number maximum of epochs is reached and the performance is minimised to a set goal;
- maximum learning time is exceeded.

Kohonen maps have proven to be very useful for classifying multi-dimensional databases dealing with nonlinear problems. The SOM algorithm is capable of extracting statistical properties present in any database input data. This includes food and food processing data.

In order to obtain satisfying results, training the network with statistically representative data among the totality of the data must be carried out. In many cases, the statistical properties of data are not clear therefore using the entire data set is necessary for good modelling. The algorithm described above is called "sequential learning" or "basic SOM". Another important learning rule is called "batch learning", which is based on the iteration of the fixed point making it much faster in terms of computation time. At each step, the BMUs for all input samples are calculated at once, and the prototype vectors are updated as follows:

$$m_i(t+1) = \frac{\sum_{j=1}^{n} h_{ij}(t)x_j}{\sum_{j=1}^{n} h_{ij}(t)} \tag{5.6}$$

5.2.3 Initialisation and parameterising of the Kohonen map

Beside the learning algorithm, there are choices to be made that can be considered as SOM settings, namely the choice of the functions $a(t)$, and $h_{ck}(t)$, the network topology, and the number of prototype vectors (and their initial state). Vector initialisation prototypes is usually carried out by one of the following methods:

Random initialisation: The prototype vectors are initialised randomly, which is often not the right policy to follow, but this policy has shown to converge to a good topographic map for a large number of learning epochs.

Linear initialisation: Prototype vectors are initialised in ascending or descending order along the x and y axes of the network. This type of initialisation depends usually main components of data samples.

Random permutation of a subset of the samples: Similar to the initialisation method, random samples are taken as vectors models.

Linear initialisation also has the advantage of being deterministic, thus reducing the character of the SOM learning algorithm. This allows easier reproduction results. We define the neighbourhood of radius r of a unit u, denoted $V(u)$, as the set of units u located on the network at a distance less than or equal to r [10, 11]. Figure 5.6 shows how the radius reaches neighbouring neurons on two different map topologies. The neighbourhood kernel $h_{ci}(t)$ can be any function which decreases with increasing the distance in the network. A typical example of a neighbourhood nucleus is derived from the Gaussian curve. Figure 5.7 shows four different neighbourhood functions to be used in a SOM. The winning neuron is selected as BMU and its influence on its

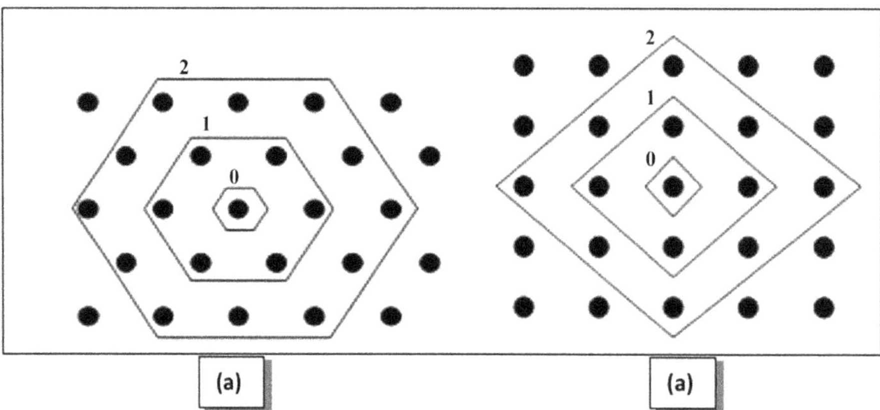

Figure 5.6: Distinct neighbourhoods (size 0, 1, and 2) of the winning neuron: (a) hexagonal network, (b) network rectangular. The innermost polygon corresponds to the order neighbourhood 0, the second to the order neighbourhood 1, and the largest corresponds to the neighbourhood.

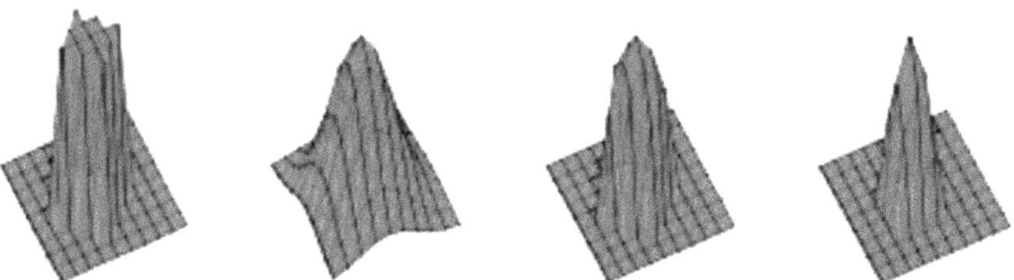

Figure 5.7: Neighbourhood functions: bubble, Gaussian, Gaussian section, and Epanechicov.

neighbours is determined by the following neighbourhood functions:

$$h_{ci}(t) = e^{-\frac{\delta_{ci}^2}{2r(t)^2}} \tag{5.7}$$

An example of SOM can be given using many libraries in many languages, one can cite the *Somtoolbox* library in Matlab [12]. The user can freely set all the above parameters, but in order to reduce any efforts, default values are provided and given in [15], as follows:

- The number of neurons in the topological layer can be defined approximately by the equation: $m = 5\sqrt{n}$, where n is the number of data samples.
- The map's default shape is a rectangular sheet with a hexagonal lattice. The length/width ratio is determined according to the ratio calculated between the two larger proper values of the covariance matrix.
- The default neighbourhood function is the Gaussian function: $h_{ci}(t) = e^{-\frac{\delta_{ci}^2}{2r(t)^2}}$, where δ_{ci} is the distance between nodes c and i on the map, and $r(t)$ is the neighbourhood radius at time t.
- The learning radius and the learning rate are monotonically functions decreasing in time. The initial radius depends on the map size, but the final radius will remain 1. The learning rate starts from 0.5 and ends, or approach, zero.
- The learning length is measured in epochs: The number of epochs is directly proportional to the ratio between the number of maps units and the number of data samples.

The SOM algorithm is very robust when it comes to choosing initialisation parameters. Thus, final results are almost often identical for different choices of functions and parameters discussed above.

5.2.4 Visualisation of Kohonen maps results

Once the SOM learning process is complete, the next step is results post-processing and visualisation. This step is particularly important, as intuitive and meaningful visualisations are in fact one of the most attractive attributes of the SOM. Although artificial neural networks are generally very difficult to visualise, SOM is a remarkable exception. This is one of the reasons making SOMs popular. The visualisation performance of SOM is mainly due to its ability to reconcile a set of data and represent them in two (or three) dimensions [13, 14]. In the visualisation techniques of clusters and variables, the SOM prototype vectors are considered a sample representing the data. The original data are replaced by a set smaller ones where the effect of noise and outliers is reduced. Assuming that properties observed when viewing prototypes are also observed for the original data, some caution is required before drawing ambitious conclusions based on SOM visualisations. In what follows, visualisation methods of the most used Kohonen networks are presented [9].

Techniques for visualising the form and structure of classes in a cloud of points are usually based on vector projections. As the shape of the SOM grid is predefined, this projection is not really useful as it is. Therefore, the prototype vector map must be projected onto a smaller space. increasingly, physical coordinates, and colour coding techniques are used for the visualisation of clusters [5, 16]. However, the most common technique used to visualise clusters obtained by a SOM is the distance matrix. Here, the distances between each unit i and the units which are in its vicinity N_i are calculated as follows:

$$D_i = \|m_i - m_j\| j \in N_i, \quad j \neq i \tag{5.8}$$

The distances, or the average distances, for each map unit are usually viewed using colours, where other visualisation techniques are possible [5, 16]. Let us take, for example, a dataset composed of three random attributes. As the data (and, therefore, the map prototypes) are three-dimensional, they can be plotted directly on a graph. In Figure 5.8, the data objects are plotted with the blue character 'O' and the prototype vectors of the Kohonen's map are drawn with the black '+' character.

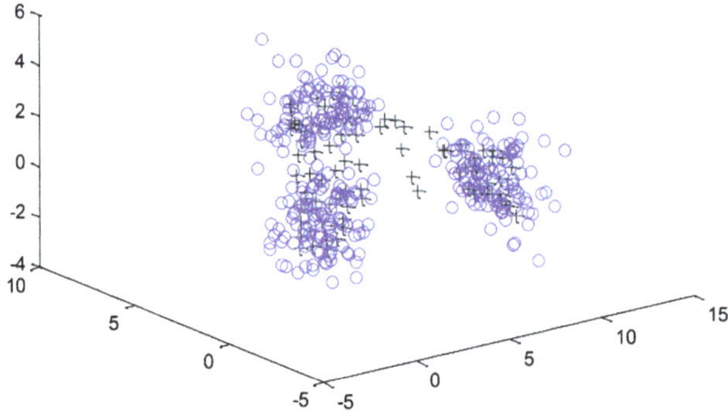

Figure 5.8: The distribution of data on the Kohonen map prototype vectors.

By direct visualisation, it is very easy to identify the distribution of data, and the way in which the prototype vectors are positioned. We can easily notice here three well-separated data groups, as well as prototype vectors between groups, but actually no data is found in these locations. Map units corresponding to these prototype vectors are referred to as "dead" or interpolation map units. Ultsch's unified distance matrix [17], called U-matrix, allows viewing all the distances of each unit from its neighbours. This is made possible thanks to the regular structure of the Kohonen map, so it is easy to position a single visual marker between each Kohonen map's unit and each of its neighbouring units. The high values on the colour bar mean large distance measurements between neighbouring units in the map and, therefore, indicate the boundaries between the clusters. A cluster, on the U-matrix map, is usually represented by a uniform region, with very small distance values between the units that compose it (refer to the colour bar to see which colours mean higher values). The U-matrix card shown in Figure 5.9 for a dataset consisting of three random attributes seems to be three clusters.

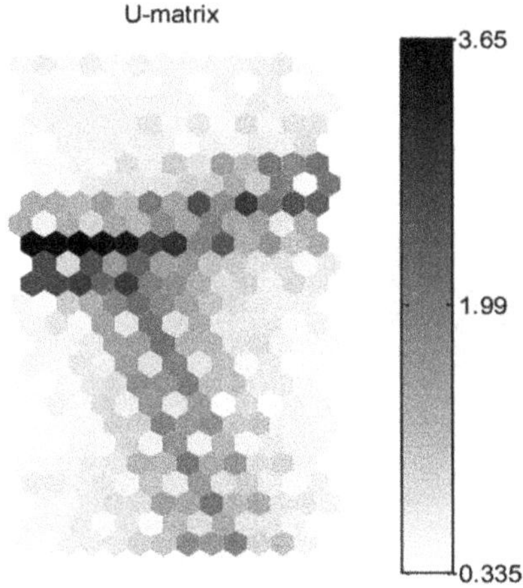

Figure 5.9: U-matrix map of all data.

The unified distance matrix contains the distances between neighbouring units on the map, as well as the average distance of each unit on the map from its neighbours. These average distances corresponding to each unit on the map can be easily extracted. The result is a mean distance matrix commonly called D-matrix. A similar technique is to assign colours to map units such as similar map units (close in terms of distance) will have similar colours. Four sub-figures representing the methods of viewing clusters previously discussed are shown in Figure 5.10: the U-matrix, the distance matrix mean (with grayscale), the mean distance matrix (with the unit size of the map), and the colour-coded matrix.

5.3 Case study: clustering of substrate methane production using Kohonen maps

Production of solid waste was world widely increased in the last decade, leading to several environmental issues. Anaerobic co-digestion, which is the anaerobic biodegradation of a mixture of two or more solid or liquid organic wastes, represents an interesting option and robust technology to reduce the waste volume and

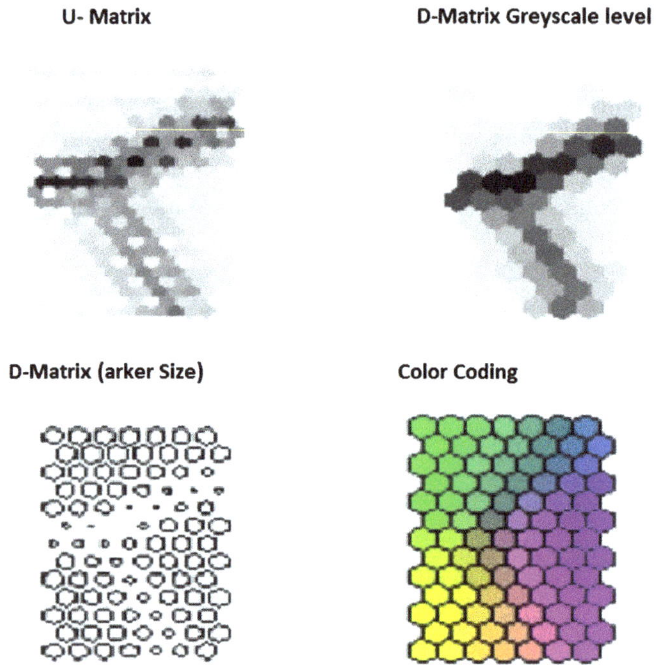

Figure 5.10: U-matrix and D-matrix map of all data and cluster results visualisation.

leading to produce biogas, mainly composed of methane/carbon dioxide which can be valorised as a renewable energy source. Such waste—as vegetables, grease, meat, and industrial by-products—has the particularity to be composed of molecules possessing different biodegradable capability. The main challenge in co-digestion is to predict the methane potential obtained when a given mixture of different wastes—or substrates—is processed. Using a database containing the quantities of methane produced over time from each substrate in batch mode, the issue is then to estimate the production of mixtures of these single substrates. A Kohonen map is used here to cluster a list of food product waste according to their ability to produce biogas. The results presented here are given in more details by the author in [18], and interested readers should refer to it for further explanations.

5.3.1 Clustering substrate by their biogas production capabilities

Methane production

The Anaerobic Digestion (AD) process is either carried out in batch (no input, no output), fed-batch (no output, a varying volume reactor) or continuous mode (output = input ≠ 0). Batch tests (easier and faster to realise than continuous ones) are usually used to study biodegradability of a waste. A batch operation consists in incubating a known amount of waste with an anaerobic microbial ecosystem and then measuring methane production over time [19].

Clustering substrate

According to methane production capabilities, basically, batch mode is used to assess Biochemical Methane Potential (BMP), which is the final volume of methane produced divided by the quantity of substrate added in term of volatile solids (VS) (mL CH4/gVSadded). Furthermore, degradation kinetics can also be identified using dynamical data obtained from batch reactors. In the last few years, fractionation of the organic matter was used in order to give more accuracy for the kinetics assessment.

In fact, exploring the evolution of methane produced over time, models the degradation solid waste in landfill conditions. Vavilin et al. in 2008 [20] reported that ideally any waste should be divided into two fractions having different kinetic rates: readily (or rapidly) biodegradable and recalcitrant (slowly) biodegradable waste. Recently, [21] fractions the organic matter into rapidly and slowly biodegradable fractions using the methane production rates obtained in batch mode. The identified kinetics were used later to calibrate an Anaerobic Digestion Modell of ADM1 type, in order to model hydrolysis of soil. Afterward, Kouas et al. in 2017 [22], proposed a new fractionation of the organic matter into three sub-fractions for 50 solid waste and a data base was established including kinetic BMPs values.

This database can be used to calibrate model optimise anaerobic digestion processes by plant operators. Subsequently, a clustering step can help to better classify the substrates and optimise the digesters feeding.

5.3.2 Available data and experimental results

Fermentation process: feedstock

In this study, the data acquired in [22], are used. In practice, 50 different solid substrates, arbitrarily divided into 9 main categories accordingly to their origin, were characterised in successive batches. The nine defined groups are:

- Fruit and vegetables: grape, peach, apple, mango, orange, banana, green cabbage, pineapple, potato, lettuce, carrot, tomato (2 varieties), cauliflower, zucchini, chayote;
- Other plant products: napier grass, grass cuttings, wheat straw;
- Vegetable by-products coming from agri-food processes: grape marc, sunflower meal, coconut meal, rape meal, beet pulp;
- Cooked based cereal products: pasta, French bread, rice;
- Animal based products: ground beef, pork fat, coalfish;
- Animal manure: cattle (3 batches), pig, chicken;

- Products obtained from the refining of vegetable oil: used winterisation earth, 2 tank sediments from the storage of rape and sunflower oils, 2 soap stocks from the refining of sunflower and rape seed oils, 2 deodorising condensates from sunflower an skimmings of aero flotation of the effluents, 1 gum from physical refining, 1 pure sunflower oil;
- Sludge obtained from domestic waste water treatment: one coming from anaerobic lagoon, and the other from a WTP operated at a medium organic loading rate (mix of primary and secondary sludge);
- Miscellaneous products: dry food pellets for guinea pigs and micro algae.

Each solid residue was characterised by:

Measuring the concentration of Total Solids (TS) and Volatile Solids (VS) permitted to characterise each solid residue. It is necessary, however, to crush all substrates, mix them, and store them at −20 °C before usage.

The biogas produced from the many batch processes are collected in a bag in order to determine the percentage of methane; hence, measuring the total methane produced from the batch.

Modelling of methane production

The volume dynamics of methane produced by a batch were used to fractionate the organic matter of the substrate into *i* sub-fractions based on their decreasing degradation rates. The following hypotheses were considered: the degradation of the different compartments of the organic matter starts straight away and simultaneously after the feeding; the degradation kinetics of each sub-fraction is constant and follows a zero-order equation ($\frac{\partial S_i}{dt} = -ki$). The entire substrate was noted S and its different subfractions were noted S_i indexed by their compartment *i*. Here, the quantity of substrate was measured in mL CH4, i. e., the volume of methane produced from a given quantity of added substrate. Thus, S_i at $t = 0$ ($S_i(0)$) is the maximum methane volume being produced from the degradation of a compartment *i* from S_i at time t ($S_i(t)$) is the volume of methane still to produce at time t by the sub-fraction S_i. The degradation rates k_i were expressed in mLCH4h-1. ($\mathrm{Vol}_i(t)$), the volume of methane produced at time t from each compartment was measured using equation (5.9).

$$\mathrm{Vol}_i(t) = \min(k_i * t, S_i(0)) \quad \text{with } i \in [1,1,3] \tag{5.9}$$

Using system identification, 6 parameters were identified for each substrate: k_i and $S_i(0)$ for $i = 1\ldots3$. The objective of the present application is to be able to classify the 50 substrates based on these 6 parameters in order to give clearer information for researchers and plant operators about the choice of inputs (substrates), to add in the digesters, considering their methane production capabilities and degradation kinetics.

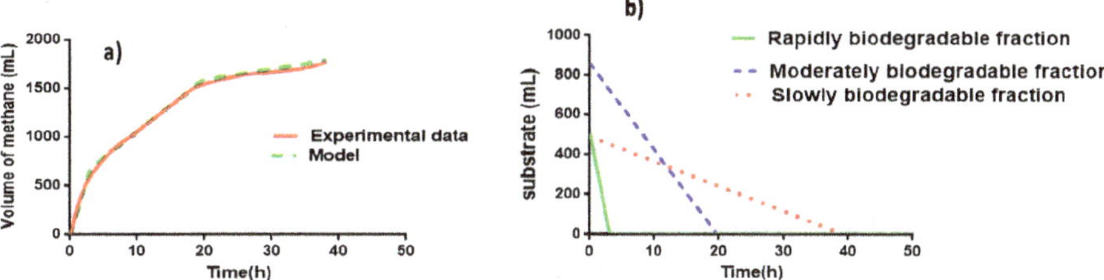

Figure 5.11: Analysis of kinetics parameters ($S_i(0)$, k_i) from the simulation of the carrot substrate batch curve: (a) the model and the experimental result in batch mode, (b) the rapidly, moderately and the slowly biodegradable fractions evolution over time.

Figure 5.11 shows the overall assessment of kinetics parameters, simulating the batch curve for carrot substrate as an example, detailing:

(a) The experimental result in batch mode obtained by the appropriate model.
(b) The rapidly, moderately, and the slowly biodegradable fractions evolution over time.

Final constructed dataset

Data on methane production are provided for 55 substrate in 50 different sheets, giving the evolution of methane production with time. Even though, the sampling time is identical for all substrates; some have longer response times than others, reaching saturation later in time. Figure 5.12 shows the methane production over time for "apples" and "sunflower sediment", respectively, left and right of the figure.

Data are compiled into one file in matrix form saturated substrates are completed with a value of 10^2 to obtain a filled homogeneous data matrix of (50×8477).

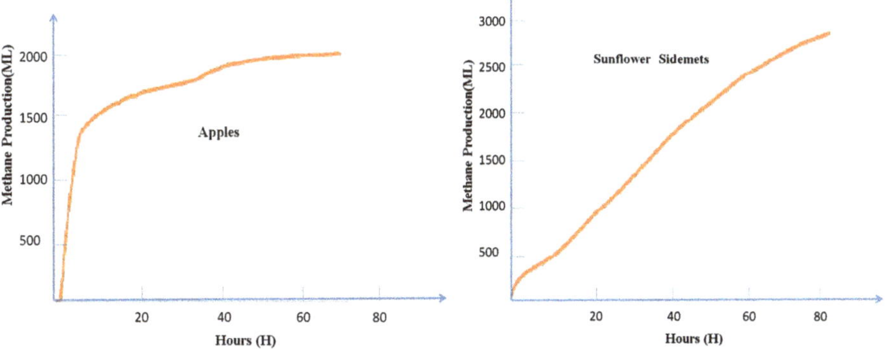

Figure 5.12: Methane substrate emission for apples and sunflower sediments.

5.3.3 Two stages Kohonen map approach for substrate methane production clustering

SOM clustering may result in a large number of prototype vectors, especially when dealing with highly multi-dimensional time series applications, and only one classification level can then be revealing.

A higher or multi-level classification is then interesting as it may provide more detailed quality analysis and less compresses the dataset when summarising all substrate production by representatives of a small class number [23]. It also can be prove difficult attributing some units of the input vector to a given cluster revealed by the map. Most problems are then encountered in the determination of cluster's borders where a clear distinction between two clusters is hardly possible. In this case, a second clustering stage proves useful to remove ambiguity and helps validate the SOM results.

The two level clustering approach used in this section and is depicted in Figure 5.13. The first abstraction level permits the creation of a set of prototypes using the Kohonen map. The obtained prototypes are then clustered in the second abstraction level using another clustering paradigm, in this case the X-means clustering algorithm.

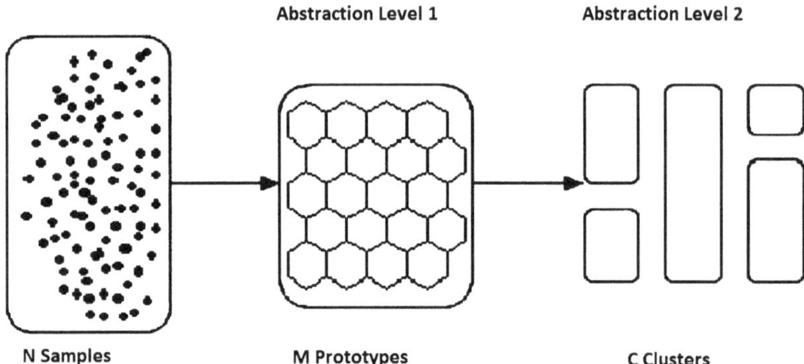

Figure 5.13: First abstraction level is obtained by creating a set of prototypes vectors using the SOM. Clustering of the SOM creates the second abstraction level.

Kohonen map results

Using a bi-dimensional map of hexagonal form Kohonen map with the following values selected empirically after several trials:

- Using the heuristic: $m = 5\sqrt{N}$, where N is the number of substrates, obtaining $m = 8$.
- A Kohonen map of a size 8×8 is then considered.
- Number of training epochs set to 200.
- A Gaussian decreasing neighbouring function with a radius $r = 3$.
- A decreasing learning rate starting at 0.5.

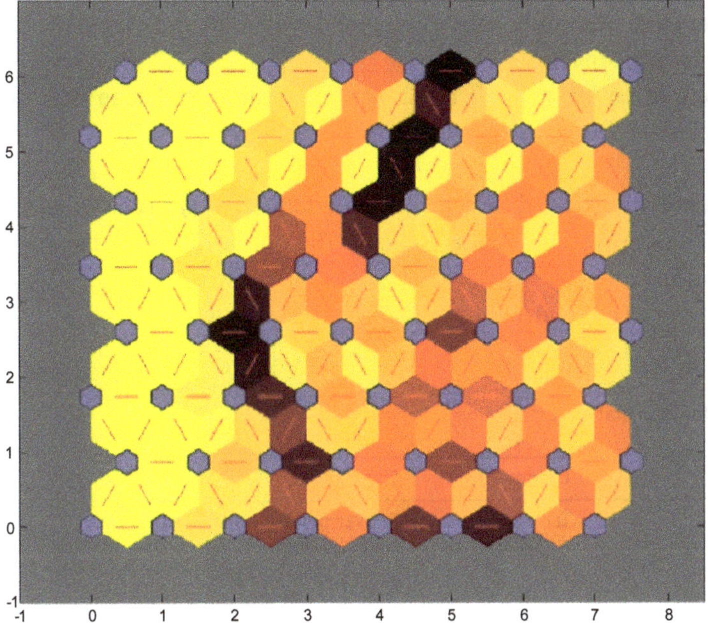

Figure 5.14: Kohonen map colour coded weights after training.

The results are visually shown after training the Kohonen map in Figure 5.14. The neurons are depicted using blue hexagons, where the red lines linking neighbouring neurons. Colours within the regions containing red lines indicate distances between neurons.

Darker colours represent larger distances, where lighter ones represent shorter distances.

A dark segment divides clearly the left region of the map from the centre. Another segment, not as clear, divides the lower right region from the centre. Thus visually three groups of substrates may be considered.

X-means results

X-means is an extension of K-means (see Section 4.2.1), where centres try to get divided within a same region. The decision between off springs and each centre is performed, comparing the obtained Information Criterion (BIC) values of both structures [23]. The main advantage is that there is no need to set in advance the number of cluster, the approach needs just to specify a cluster intervals (minimum and maximum cluster number). The interval is obtained from the a priori clustering using Kohonon maps. In this case, two scenarios with a minimum number of clusters set in turns to 2 and 3 each time defining each time the maximum number of clusters being the number of substrates (i. e., 50 clusters). The obtained clustering results are for each case:

Defining 2 Clusters as a minimum after training and visualisation, X-map settles to a number of clusters equal to 3.

Defining 3 Clusters as a minimum after training and visualisation, X-map settles to a number of clusters equal to 5.

Regrouping substrates in identified clusters is then performed for each clustering solution (3 and 5 clusters) using conventional K-means. The final results are summarised in Table 5.1 and Table 5.2.

The following can be concluded from Table 5.1.

Table 5.1: Three clusters substrates classification.

Cluster 1	Cluster 2	Cluster 3
Straw, orange, Bovine manure 2, 2, apple, Sheep manure, Gum, bovine manure 1, Chayotte Andres 1, Banane Andres 1, Ananas Andres 1, Mangue 1, Andres 1, Grass, Napier, Mixture mango chou flower, Table oil, Tomate2, Cauliflower, fish, Beef, Naskeo algae, Biodiesel waste, Coconut cake, Grape Marc, Tomato 1, grapes, carrot, Salad, PdeT 1, Cabbage peach, bread, Cooked pastas, cooked rice, Sludge, from aerobic lagoon, Dry food pellets guinea pigs, algues2, sunflower meal, Rape meal, beet pulp, soapstock grape oil, soapstock (rape oil), condenstae (suflower oil), Skimming of aeroflottaion 1, Skimming of aeroflottaion 2, Sunflower, table oil, Sludge from WWTP, pig manure.	Tank sidement, suflower meal	Graisse porc tank sediment, sunflower oil, earth sunf soapstock, Condensate (palm oil), Chicken excrements. Used winterisation, sunflower oil, Courgette

Table 5.2: Five clusters substrates classification.

Cluster 1	Cluster 2	Cluster 3	Cluster 4	Cluster 5
Straw, orange, Courgette2, apple Chayotte Andres 1, Banana Andres 1 Mangue 1 Andres 1, Tomate2, fish, Beef Biodiesel waste, Tomato 1, grapes, Carrot, peach, soapstock grape oil, condenstae (sunf oil), Skimming of 1, aeroflottaion Tank sediment, Cooked rice, sunflower meal, Sludge from WWTP	Tank sediment, Sunflower meal, Naskeo	Sheep manure, Bovine Manure 1, Grass, Napier, seaweed, Sludge from aerobic lagoon, algues2, sunflower meal, Skimming of aeroflottaion 2, Pig manure	Bovine manure 2, Ananas Andres 1, sidement, Mango, cauliflower mixture, Table oil, Cauliflower, Coconut, crab, grape, Marc, Cabbage, Bread, Cooked pastas, Dry food pellets, guinea pigs, Rape meal, Beet pulp, soapstock (rape oil), Sunflower table oil, Salad, PdeT 1	Pork fat, Tank sunflower oil, Used winterisation, earth sunf soapstock (sunflower oil), Condensate (palm oil), gum, chicken Excrements

Cluster 1: Regroups substrates representing vegetal organic waste product, having similar saturation curves.

Cluster 2: This cluster is quite unique and contains only one substrate totally different in term of methane production than the rest. Note that the cluster is the same even for the 5 cluster choice (Table 5.2).

Cluster 3: Regroups substrates from animal waste origins, as well as natural fertilizers. These substrates have similar methane production curves and saturates in a similar manner.

5.3.4 Conclusions

Methane production analysis using Kohonen maps as a first level clustering layer allowed us to obtain a rough visual identification of the different existing substrate clusters. The X-means algorithm coming as a complement for better class clustering in a second clustering layer, permitted to define clearer frontiers between clusters. Using a two-stage clustering strategy proved more efficient than a single-layer clustering approach involving only SOM or X-means algorithms alone. The obtained classification is also found to be more compact as it merges neighbouring clusters into one, depending on the initial cluster choice. Different clusters have been identified (3 and 5 depending on minimum cluster initialisation) with a clear borders definition of methane production capabilities, also providing a comprehensive analysis for different cluster's constituents in terms of substrate nature and origin. The results obtained may then be used to design prediction models systems according to the number and the nature of each cluster for methane production; mixes of substrate and their effect on production may then be simulated.

Bibliography

[1] Dreyfus, G., Martinez, J. M., Samuelides, M., Gordon, M. B., Badran, F., Thiria, S., and Hérault, L. (2004) Réseaux de neurons: méthodologies et applications. édition EYROLLES Avril 2004.

[2] Kohonen, T. (1990) The self-organizing map. Proceedings of the IEEE, 78, 1464–1480.

[3] El Golli, A., Conan-Guez, B., and Rossi, F. (2004) A self-organizing map for dissimilarity data?, in Proc of IFCS'. Elminir, H., Abdel-Galil, H., (2006) Estimation of air pollutant concentrations from meteorological parameters using artificial neural network. Electr. Eng. 57 (2), 105–110.

[4] Boinee, P. (2006) Insights into machine learning: data clustering and classification algorithms for astrophysical experiments (Doctoral dissertation, Ph. D thesis, Dept. of Math and Computer Science, 2006, Univ of Udine–Italy).

[5] Vesanto, J. (1999) SOM-based data visualization methods. Intelligent Data Analysis, 3(2), 111–126.

[6] Himberg, J. (2000) A SOM based cluster visualization and its application for false coloring, in IJCNN 2000, Proceedings of the IEEE-INNS-ENNS International Joint Conference on Neural Networks, Vol 3, pp. 587–592.

[7] Reljin, I. S., Reljin, B. D., and Jovanovic, G. (2003) Clustering and mapping spatial-temporal datasets using SOM neural networks. Journal of Automatic Control, 13(1), 55–60.

[8] Turias, I. J., Gonzalez, F. J., Martín, M. L., and Galindo, P. L. (2006) A competitive neural network approach for meteorological situation clustering. Atmospheric Environment, 40(3), 532–541.

[9] Pölzlbauer, G. (2004) Application of self-organizing maps to a political dataset. Master Thesis, Vienna University of Technology.

[10] Rousset P. (1999) Applications des algorithmes d'auto-organisation à la classification et à la prévision, thèse de doctorat, Univ. Paris I.

[11] Lemaire V. (2006) Cartes auto-organisatrices pour l'analyse de données, in Proc. Confèrence en Recherche d'Information et Application (CORIA).

[12] Vesanto J., Himberg, J., Alhoniemi, E., and Parhankangas, J. (1999) Self-organizing map in Matlab: the SOM Toolbox, in Proc. of the Matlab DSP conference, Vol. 99, pp. 16–17.

[13] Simula, O., Vesanto, J., Alhoniemi, E., and Hollmén, J. (1999) Analysis and modeling of complex systems using the selforganizing map, in Neuro-Fuzzy Techniques for Intelligent Information Systems, pp. 3–22.

[14] Kaski, S. (1997) Data exploration using self-organizing maps, in Acta Polytechnica Scandinavica: Mathematics, Computing and Management in Engineering Series, 82.

[15] Vesanto, J., and Alhoniemi, E. (2000) Clustering of the self-organizing map. IEEE Transactions on Neural Networks, 11(3), 586–600.

[16] Vesanto, J. (2002) Data Exploration Process Based on the Self-Organizing Map. Helsinki University of Technology.

[17] Ultsch, A. and Siemon, H. (1990) Kohonen's self-organizing feature maps for exploratory data analysis, in Proc. INNC'90, Int. Neural Network Conf., pp. 305–308, Dordrecht, Netherlands.
[18] Kouas, M., Khadir, M. T., Meddour, A. and Harmand, J. Clustering of Substrate Methane Production Using Kohonen Self-Organising Feature Maps, Conference: The International Conference on Control, Automation and Diagnosis 2018 (ICCAD'18).
[19] Hansen, T. L., Schmidt, J. E., Angelidaki, I., Marca, E., Jansen, J. L. C., Mosbaek, H., and Christensen, T. H. (2004) Method for determination of methane potentials of solid organic waste. Waste Management. 24, 393–400.
[20] Vavilin, V. A., Fernandez, B., Palatsi, J., and Flotats, X. (2008) Hydrolysis kinetics inanaerobic degradation of particulate organic material: an overview. Waste Management. 28, 939–951.
[21] Garcia-Gen, S., Sousbie, P., Rangaraj, G., Lema, J. M., Rodríguez, J., Steyer, J.-P., and Torrijos, M. (2015) Kinetic modelling of anaerobic hydrolysis of solid wastes including disintegration processes. Waste Management. 35, 96–104.
[22] Kouas, M., Torrijos, M., Sousbie, S., Steyer, J. P., Sayadi, S., and Harmand, J. (2017) Robust assessment of both biochemical methane potential and degradation kinetics of solid residues in successive batches. Journal of Waste Management, 70, 59–70.
[23] Kato, S., Horiuchi, T., and Itoh, Y. (2009) A study on clustering method by self-organizing map and information criteria, in International Conference on Neural Information Processing, Bangkok, Thailand, pp. 874–881.

6 Deep artificial neural networks

6.1 Introduction

Deep Learning (DL) describes a set of calculation models composed of multiple data processing layers, which allow learning by representing this data by several abstraction levels. Assuming a large number of training data, these models discover recurring structures by automatically refining their internal parameters via a learning algorithm (such as back-propagation). DL approaches are part of the family of so-called Automatic learning computer techniques (or machine learning, ML), which is an area of artificial intelligence.

As described in Chapter 2, Deep Artificial networks are the new trend in machine learning, and its concept has been firstly introduced in the LeNet CNN network in [1]. However, to understand the causes of this growing success, we have to wait until the ImageNet competition of 2012, which is considered as the turning point that will definitively popularise Deep Learning (DL) in the general public. ImageNet is an image classification competition of taking place every year since 2010, which provides a database about 1.2 million images labeled in 1000 different classes and which confront research teams from around the world. In the 2012 edition, a team succeeds to everyone's surprise, halving the rate of image classification errors. This team, the only one to use a DL architecture, is developing a network (AlexNet) to nine layers composed of 650,000 neurons and 60 million parameters [4]; Alexnet being largely inspired from LeNet [1]. In the next two editions of this challenge, all teams use DL approaches, and the algorithm error rate is now less than 5 %, equivalent to that of a human observer [5].

The success of the network developed by [4] rests on its architecture, on the power of calculations, on GPU graphics cards and on the availability of a lot of training data. It is the conjunction of these three factors that have enabled them to achieve spectacular results, and which today places DL in the lead AI approaches. The calculation on GPU makes it possible to manage in an acceptable time (a few hours) the memory required for the learning and adjustment processes billions of weights of contemporary networks. In order for the generalisation capacities of a DL network to be efficient, it is imperative that the number of training data is large enough to be significant. This is the case today for image recognition with labelled image banks available through the internet, however less present in food application technology. Hence fewer application of DL in food technology (Section 2.5).

6.1.1 Definition of deep learning

Deep learning is based on what has been called, by analogy of "networks of artificial neurons", made up of thousands of units ("neurons") that each perform small simple

https://doi.org/10.1515/9783110646054-006

operations. The results of a first layer of "neurons" serve as input to the calculations of a second layer and so on.

LeCun has defined deep DL as follows: "Deep learning allows computer models composed of several processing layers to learn data representations with multiple levels of abstraction" [2].

6.1.2 Principles of deep learning

The principles of DL are based on the extension of a single neuron to form the well-known multi-layered perceptron, as a deep network is just a MLP with an important number of layers.

Equation (6.1) expresses the stack of three layers of perceptrons to form a two hidden layer MLP. Beyond three layers, we may talk of DL and Deep networks shown in Figure 6.1. It should be noted that the number of perceptrons per layer is variable and depends on the data processed as well as the type of architecture used [6]:

$$y = \sigma(W_{3\sigma}(W_{2\sigma}(W_{1\sigma}))) \tag{6.1}$$

The objective remains, as for MLPs, the search for appropriate weights vectors W_i. The main difference is the way to train the network in order to find the appropriate weights as the back-propagation algorithm lacks efficiency when the number of hidden layers increases.

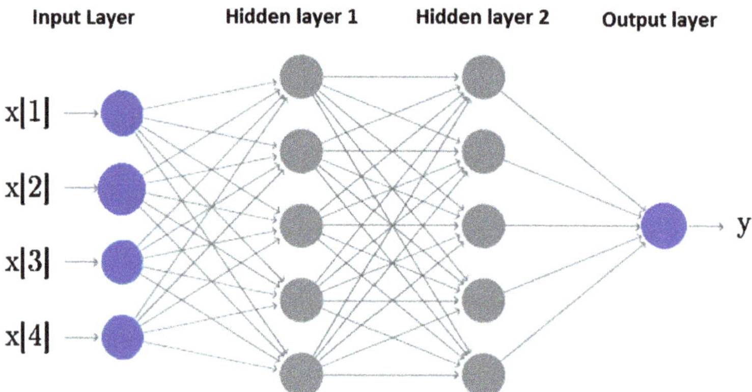

Figure 6.1: Two hidden layer MLP: the basic architecture of a DN.

6.2 Architectures of deep neural networks

There are many deep network architectures available that can be used depending on applications or specific learning data. Several classifications are possible, which are detailed in [7]. In what follows, the classification of the mostly used architectures presented in [3] is described.

Figure 6.2 depicts the general DNN architectures used for both supervised and unsupervised learning. The six most important architectures are presented and described in more details in the following sections.

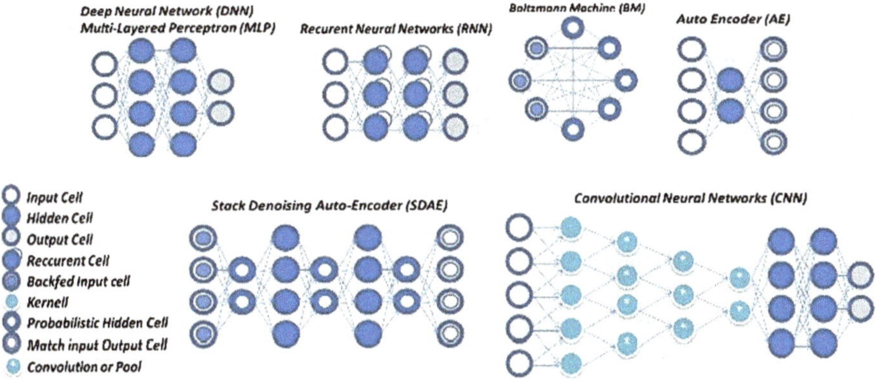

Figure 6.2: DNN architectures.

6.2.1 Recurrent neural network (RNN)

Recurrent neural networks are adapted to process time-dependent information (or time series data), such as speech or video processing. Contrary to other types of deep networks, the idea here is to remember information previously processed to help the network predict the following data. The output y_t of the network is therefore a function not only of the input x_t at a time t but also inputs x_{t-i} at times $t - i$. Among RNNs are the so-called Long Short-TermMemory (LSTM) models and Gated Recurrent Unit (GRU).

6.2.2 Auto-Encoder (AE)

Auto encoder is a network with a hidden layer of dimension lower than the entry. The hidden layer plays the role of an encoder, in order to identify a dominant latent structure of reduced size by compared to the input signal. The input layer neurons are fully connected to the hidden layer, which are all connected to the output layer in a feedforward manner. When the encoding layer is used as input to another AE, we speak of stacking of AE (Stacked Auto-Encoder, SAE), which generates several levels of abstraction. Several variants have been proposed: Stacked Denoising (SDAE), Sparse (SAE) and Variational (VAE). An advantage of stacking AE is that it can be used in the framework for unsupervised learning, requiring no labelled data.

6.2.3 Deep Belief Network (DBN)

Deep Belief Networks are essentially Stacked Auto Encoders (SAEs) in which the encoding layers are replaced by RBMs. Only the two deeper layers have two-way connections. The training is carried out using unsupervised learning, until reaching final network adjustments. Then the parameters are tuned is a supervised manner by adding an output classification layer network. The Deep Boltzmann Machine (DBM) may be considered as the equivalent of a bidirectional DBN, with connections between neurons of each layer.

6.2.4 Boltzmann machine

Boltzmann Machine (BM) is another DL architecture that uses unsupervised learning. The objective stays the same as for an AE, for instance: extract representations, based however, on a different statistical model. The connections between the neurons are bidirectional and, therefore, comparable to a so-called generative model that can generate new input data during learning. In a standard BM, all neurons are fully connected to each others, while only the neurons of distinct layers are, in a Restricted Boltzmann Machine (RBM).

6.2.5 Recurrent Neural Network (RNN)

Recurrent neural Networks, are deep Multi-layer Perceptron (MLP), which contains more than three hidden layers. Therefore, in such an architectures, several hidden layers in which all of the neurons in a layer i are connected to all the neurons of layer $i+1$. This is the simpler DNN architecture, and suffers from slow computational learning as it still uses the back-propagation algorithm, which complexity increases exponentially with the addition of a new layer.

6.2.6 Convolutional Neural Network (CNN)

Convolutional Neural Network is the DL architecture used for the first time in [4] and mainly used nowadays in image processing. This type of DNN is of paramount importance and will be described in details in Section 6.6.

6.3 Bolzmann Machine (BM)

In 1985, Geoffrey Hinton, a professor at the University of Toronto, introduced Boltzmann Machines (BMs) to the world. BM is a generative unsupervised neural network model, which involve learning a probability distribution from an original dataset. The learning phase is afterwards used to infer knowledge about unknown data.

The Boltzmann machine is a neural network with an input layer and one or several hidden layers. The specificity of BM is the ability of making stochastic decisions about whether to turn on or off based on the data we feed during training, minimising the cost function during training. BM allows then discovering interesting features present in data and impossible to see due to high dimensionality. The discovered features help model the complex underlying relationships and patterns present in the data.

The Boltzmann machine possesses an input layer (referred to as the visible layer) and one or several hidden layers (referred to as the hidden layers). In BM, neurons are fully connected, not only to other neurons in other layers but also to neurons within the same layer, Figure 6.4. The connections nature are bi-directional, where visible neurons and hidden neurons are connected to each other, generating data. Neurons generate information regardless they are hidden or visible. For the Boltzmann machine, all neurons are similar, and does not discriminate between hidden and visible neurons.

BM has fixed weights, hence the training weight update proceeding used for other ANN types (such as back-propagation algorithm for MLPs) is not applicable. However, setting the network weights and finding a Consensus Function (CF) became the new target.

Considering a BM that has a set of units U_i and U_j that has bi-directional connections, let us consider:
- fixed weights w_{ij},
- $w_{ij} \neq 0$ if U_i and U_j are connected,
- a symmetry between weighted interconnection, i. e., $w_{ij} = w_{ji}$,
- there are self-connections between unit, i. e., w_{ii} = exists,
- the state u_i would be either 1 or 0 for any unit U_i.

The main objective of the Boltzmann machine is to maximise the consensus objective function CF, becoming then, the main goal of the BM and is formulated by the following relation:

$$\mathrm{CF} = \sum_i \sum_{j<i} w_{ij} u_i u_j \tag{6.2}$$

The state changes of 1 to 0 or 0 to 1 induce a change in the consensus function given by the following equation:

$$\Delta\,\mathrm{CF} = (1 - 2u_i)\left(w_{ij} + \sum_{j \neq i} w_{ij} u_i \right) \tag{6.3}$$

Here, u_i is the current state of U_i.

The variation in the coefficient $(1 - 2u_i)$ is given by the following relation:

$$(1 - 2u_i) = \begin{cases} +1, & U_i \text{ is currently off} \\ -1, & U_i \text{ is currently on} \end{cases}$$

In general, no state changes on the unit U_i occurs; however, in the case it happens, the change information reside in the unit.

Probability of the network to accept the unit state change is given by the following relation:

$$AF(i, T) = \frac{1}{1 + e^{-\frac{\Delta\,\mathrm{CF}(i)}{T}}} \tag{6.4}$$

where T is the controlling parameter. It will decrease as CF reaches the maximum value.

6.4 Recurrent neural networks: Long Short Term Memory Networks (LSTM)

The conceptual dissimilarity that differs recurrent neural networks from the standard feedforward networks is the nature of the activation signal flow through numerous units within each layer, where the recurrent topologies allow a cyclical connections

The difference between RNNs and MLPs is structural and topological, even if the forward pass remains similar to both with the exception that activations happen at the hidden level. The activation are influenced however by both the external input of the neuron itself and the previous time-step hidden layer activations. In that manner, the RNN establishes a temporal correlation between the current state and previous inputs. When the number of Hidden layers exceeds three, we can speak about deep LSTM and deep RNNs.

6.4.1 Forward pass

LSTMs have proven to perform better than other traditional RNNs and tasks involving long time lags. The LSTM architecture allows a greater number of successful runs along with faster learning compared, for example, to Recurrent Learning (RTRL), Recurrent Cascade-Correlation, Back Propagation, and Elman Nets [8].

The main difference between LSTM and standard ANNs reside in the architecture of the hidden layer(s). Indeed, the summation unit of the traditional MLP is replaced instead by memory blocks.

Each block has three input sources and contains one or more self-connected memory cells: the input z_m^{in}, output z_m^{out} and the cell itself z_m^{φ}. Every source is squashed with

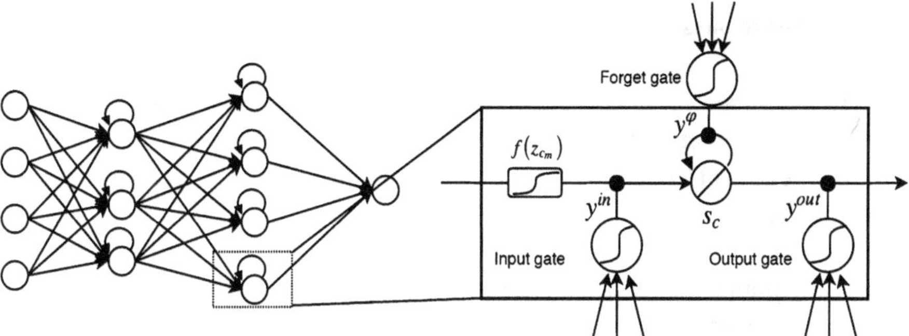

Figure 6.3: LSTM memory block with one cell.

an activation function, termed as "gate" providing continuous regulators that can perform: write, read, and reset operations for every cell. Illustration of an LSTM memory block with a single cell is provided in Figure. 6.3:

$$z_j = \sum_{i=1}^{K} w_{ji} y_i + w_{j0} \tag{6.5}$$

More precisely, this means that y_m in equation (6.5) is replaced by the following set of equations:

$$z_{c_m}(t) = \sum_{m=0}^{k} w_{c_m j} y_j(t-1) \tag{6.6}$$

where $z_{c_m}(t)$ constitutes the network input cell, firstly calculated during each forward pass as:

$$z_m^{\text{in}}(t) = \sum w_{mj}^{\text{in}} y_j(t-1); \quad y_m^{\text{in}}(t) = f_m^{\text{in}}(z_m^{\text{in}}(t)) \tag{6.7}$$

$$z_m^{\varphi}(t) = \sum w_{mj}^{\varphi} y_j(t-1); \quad y_m^{\varphi}(t) = f_m^{\varphi}(z_m^{\varphi}(t)) \tag{6.8}$$

$$z_m^{\text{out}}(t) = \sum w_{mj}^{\text{out}} y_j(t-1); \quad y_m^{\text{out}}(t) = f_m^{\text{out}}(z_m^{\text{out}}(t)) \tag{6.9}$$

where m refers to memory block with only one cell c_m; see [12] for details. Moreover, s_{c_m} indexes the cell state of the mth memory block which is updated according to equation (6.10).

$$s_{c_m}(t) = y_m^{\varphi}(t) s_{c_m}(t-1) + y_m^{\text{in}}(t) f(z_{c_m}(t)) \tag{6.10}$$

with

$$s_{c_m}(0) = 0 \tag{6.11}$$

6.4.2 Backward pass

LSTM weights update is mainly performed in the backward pass and is a fusion of BP for output neurons and gate weight changes. It may be viewed as a slight modified and truncated version of RTRL update algorithm, for input weights, input gates, and forget gates update.

The gradient is then truncated after every time-step and not according to the activation flow around the recurrent connection. The gradient is therefore altered avoiding the gradient vanishing and exploding problem [9]. For a full derivative of the algorithm, see [11].

For the output units and output gate weight changes, are computed via gradient descent obtained by standard back propagation (equation (6.12)), given in more detail in Section 3.6:

$$\Delta w_{mj}^{out}(t) = \mu \delta_m^{out}(t) y_j(t) \tag{6.12}$$

$$\delta_m^{out}(t) \doteq f'^{out}_m(z_m^{out}(t)) \Big(\sum s_{c_m}(t) \sum w_{kc_m} \delta_k(t) \Big) \tag{6.13}$$

Here, $\doteq$, represents error truncation:

$$\frac{\partial s_{c_m}(t)}{\partial w_{mj}^{in}} \doteq \frac{\partial s_{c_m}(t-1)}{\partial w_{mj}^{in}} y_m^{\varphi}(t) + f(z_{c_m}(t)) f'^{in}_m(z_m^{in}(t)) y_j(t-1) \tag{6.14}$$

$$\frac{\partial s_{c_m}(t)}{\partial w_{c_{mj}}} \doteq \frac{\partial s_{c_m}(t-1)}{\partial w_{c_{mj}}} y_m^{\varphi}(t) + f'(z_{c_m}(t)) y_m^{in}(t) y_j(t-1) \tag{6.15}$$

$$\frac{\partial s_{c_m}(t)}{\partial w_{mj}^{\varphi}} \doteq \frac{\partial s_{c_m}(t-1)}{\partial w_{mj}^{\varphi}} y_m^{\varphi}(t) + s_{c_m}(t-1) f'^{\varphi}_m(z_m^{\varphi}(t)) y_j(t-1) \tag{6.16}$$

Internal state error $e_{s_{c_m}}$ is calculated separately for each memory cell in order to calculate weights changes

$$e_{s_{c_m}}(t) \doteq y_m^{out}(t) \Big(\sum w_{kc_m} \delta_k(t) \Big) \tag{6.17}$$

6.4.3 Update process

The weights represent connections between hidden layers and inputs, where cell and the forget gates are updated using the partial derivatives from equations (6.14), (6.15), and (6.16):

$$\Delta w_{c_m}(t) = \mu e_{s_{c_m}}(t) \frac{\partial s_{c_m}(t)}{\partial w_{c_{mj}}} \tag{6.18}$$

$$\Delta w_{mj}^{\text{in}}(t) = \mu e_{s_{c_m}}(t)\frac{\partial s_{c_m}(t)}{\partial w_{mj}^{\text{in}}} \tag{6.19}$$

$$\Delta w_{mj}^{\varphi}(t) = \mu e_{s_{c_m}}(t)\frac{\partial s_{c_m}(t)}{\partial w_{mj}^{\varphi}} \tag{6.20}$$

6.5 Stack Denoising Auto-Encoder (SDAE)

6.5.1 Auto-Encoder (AE)

A single Auto-Encoder (AE) [15], may be considered as a neural network type with a topological particularity of having the same number of input and output nodes in the input and output layers regardless of hidden network configuration. The purpose of AEs is the reconstruction of its inputs instead of some target values.

The AEs hidden layer(s) activation computes an encoded version of inputs g. The transformation of input data to the value: z is then mapped back with a decoder to an input's reconstruction to g' as follows:

$$z = \sigma(Wg + b) \tag{6.21}$$

$$g' = \sigma(Wz' + b') \tag{6.22}$$

where b, b' is bias column vector, W, W_0 is the weights, and σ is the sigmoidal activation function $\frac{1}{1+e^{-v}}$.

6.5.2 Denoising Auto-Encoder (DAE)

A way to reduce target and construction errors as well as increasing performances is to stack multiple layers of AEs with the introduction of a denoising criterion at the input level used for the unsupervised task, helping obtaining a higher level representation [13].

The obtained termed, Denoising Auto-Encoder (DAE), is then trained to reconstruct a clear version of its inputs. In the training stage, each DAE hidden layer computes an encoded version of its input, while a Gaussian corruptor is applied to the inputs, adding zero mean Gaussian noise as follows:

$$\dot{x} = x + bN(0, 1) \tag{6.23}$$

Let $\dot{x}$ being the corrupted input, N denoting a normal distribution, and $b > 0$ is the noise factor.

Every DAE of a Stacked DAE configuration is trained using a mini-batch gradient descent algorithm [15] with the goal of minimising the Mean Squared Error (MSE) as expressed in equation (6.24) between the original input x and its reconstructed version

x', being x_{ij} the component i of the pattern j, N is the size of mini-batch, and M is the number of components (inputs):

$$\text{MSE} = \sum_{j=1}^{n}\sum_{i=1}^{m}(x_{ij} - x'_{ij})^2 \tag{6.24}$$

Once the weight update is performed, as described in the previous steps, they are used to initialise the feedforward neural network of each model. Only then, each neural network can be trained in a supervised way based on the difference between calculated and target values.

6.5.3 Stacking Denoising Auto-Encoder (SDAE)

Denoising Autoencoders (DAEs) can be stacked to form a deep network called Stacked Denoising Autoencoder (SDAE) by using initially a local unsupervised criterion to pre-train each layer in turn. Every layer is trained as a single DAE by minimising the error in reconstructing its input and learning to produce a useful higher-level representation from the lower-level representation output of the previous layer. Once pre-training of all layers is completed, the network goes through a second stage of training called fine-tuning, where we want to achieve a better generalisation performance by minimising the prediction error on a supervised task. At this point, the entire network is trained as would be trained a multi-layered perceptron by considering only the encoding part of each auto-encoder [13]. This stage is supervised, therefore, the target values during training are used. The main advantage of stacked denoising auto-encoders is that the usage of more hidden layers when compared to a single auto-encoder, permits that a high-dimensional input data can be reduced focusing on the most important input features. The presented deep neural network, Figure 6.4, is composed of 3 denoising auto-encoders, where the dimensions of every encoding one are respectively 50, 20, and 8 neurons each. The amount of noise added to the input data before performing the unsupervised training, has to be chosen and set by the user and is usually chosen between 20 and 60 %.

Figure 6.4 illustrates the working procedure of a stacked denoising auto-encoder composed of 3 denoising auto-encoders. The first step is the unsupervised training, where the first auto-encoder encode the corrupted input x to a smaller representation S in order to take just the more useful and significant features, and then decode it to reconstruct the input x'. The output of the first denoising auto-encoder is the code found in the hidden layer which is considered as the input layer in the next auto-encoder. At this stage we perform the same previous steps for the second and third auto-encoders. When all the denoising auto-encoders are pre-trained, we go forward to the next stage of training called fine tuning, the code part of each denoising auto-encoder is stacked to form the hidden layers of a multi-layer perceptron, keeping the same input layer as in the first step and adding an output layer with one unit.

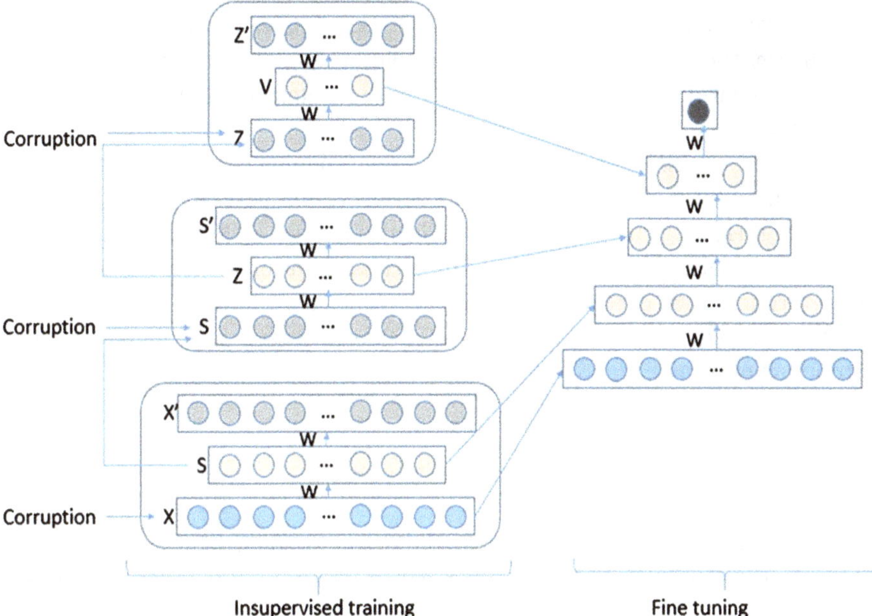

Figure 6.4: Three layer SDAE.

6.6 Convolutional Neural Network (CNN)

In this section, the different layers of convolutional neural networks, as well as their most used architectures are described. According to Yan LeCun [2], the typical architecture of a CNN is structured in a series of steps and stages. The first stages are made up of two types of layers: convolutional layers and pooling layers.

The units of a convolutional layer are organised into characteristic maps which are going to be corrected by a set of filters whose values are parameters that the network learns through the gradient descent algorithm, seen in 3.6. The result is then passed to a nonlinear function. Grouping layers permits to highlight relevant features. The last stages are structured in layers which carry out the classification of these images (or two-dimensional features) using these characteristics in separate classes.

In general, CNNs are organised into four layer types, for feature extraction and fully connected layers for classification, as follows:
- Convolution layer followed by a nonlinear ReLU activation function.
- Pooling layer.
- Standardisation layer.
- Fully connected layer.

The first part of a CNN is the convolutional part itself and works as an image feature extractor. An image (or a 2-dimensional data) are passed through a succession of filters, or convolution kernels, creating new images called convolution maps (Figure 6.3), some intermediate filters reduce the image resolution by a local maximum operation. Finally, the convolution maps are flattened and concatenated into a feature vector, called CNN code.

The obtained CNN code at the output of the convolutional part is then connected to the input of the second part, consisting of fully connected layers network (MLP). The role of this part is to combine the characteristics of the CNN code, obtained from the convolutional and pooling layers to finally classify the image. The output is a final layer with one neuron per category or class, where obtained numerical values are generally normalised between 0 and 1, of sum of ones in order to produce a probability distribution over categories or classes.

6.6.1 Training steps of a convolutional neural network

Firstly, creating a new convolutional neural network is costly in terms of expertise, material, and amount of annotated data required. The procedure starts by fixing the network architecture, including the number of layers, their sizes, and the weight matrix.

Training consists then in optimising the network weight coefficients to minimise the classification error output. This training can take several weeks for large CNNs, with many GPUs working on hundreds of thousands of annotated images. The following steps are the standard procedure to train a CNN.

Building blocks: CNN architecture is formed by a stack of processing independent layers.

The convolution layer (CONV): Three hyper-parameters allow to size the volume of the convolutional layer (also called the output volume): the “depth”, the “step”, and the “margin”.

1. Layer’s depth: is the number of convolution nuclei (or number of neurons associated with the same receptive field or layer).
2. The step value: controls the overlap of the receiving fields. The smaller is the step value, the more the receptive fields overlap and greater will be the exit volume.
3. Margin set to 0, or “zero padding”: sometimes it is convenient to put zeros at the border of the entry volume. The size of this zero-padding is the third hyper-parameter. This margin allows to control the output spatial volume dimension. In some cases, it is desirable to keep the same area and the entry volume size equal [14].

Pooling layer (POOL): Another important CNN concept is pooling, which is a form of image sub-sampling. The input image is cut into a series of nonoverlapping rectangles of n pixels for each side. Each rectangle can be seen as a tile. The output signal of each tile is defined according to the values taken by the different pixels of the tile. Pooling reduces the spatial size of an intermediate image, thereby reducing the quantity of parameters and calculation in the network. It is therefore frequent to periodically insert a pooling layer between two convolutional successive layers of a given CNN architecture in order to control overfitting. The pooling operation also created a form of invariance by translation.

The pooling layer works independently on each slice of the depth of the entrance and resizes it only at the surface level. The most common form is a pool layer with tiles of size 2×2 (width/height) and as an output value, the maximum input value (see Figure 6.5). We speak in this case of "Max-Pool 2×2" (compression by a factor of 4). It is possible to use other pooling functions than the maximum. An average pooling may also be used, where the output is the average of the input values batch, "L2-norm pooling". In fact, even if initially pre-pooling was often used, it turned out that max-pooling was more effective because it increases the importance of stronger activations. In other circumstances, stochastic pooling may also be used.

Pooling, in general, allows big gains in computing power. However, due of the aggressive reduction in the size of the representation (and, therefore, of the loss associated information), the current trend is to use small filters (2 × 2 type). It is also possible to avoid the pooling layer but this implies a higher risk of over-training [16].

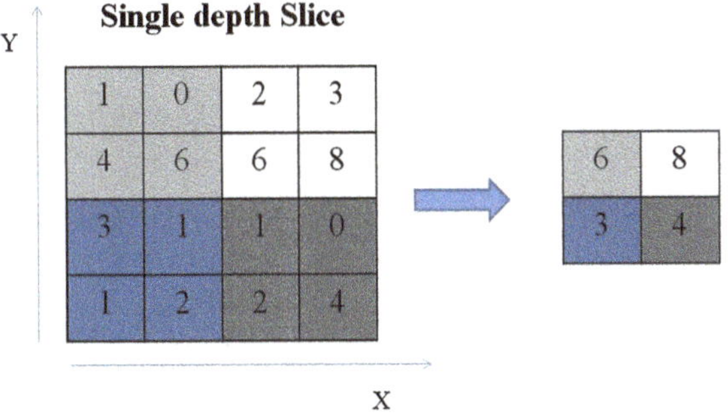

Figure 6.5: 2 × 2 Max pooling example.

Correction layer (ReLU): It is often possible to improve the treatment effectiveness by intercalating between the processing layers, a layer operating a mathematical function (activation function) on the output signals. Among the most popular, and similar to the MLP, one may cite:

1. The hyperbolic tangent correction $f(x) = \tanh(x)$.
2. The correction by the saturated hyperbolic tangent: $f(x) = |\tanh(x)|$.
3. Correction by the sigmoid function.

Often the Relu function, or correction function, is preferable, and is found to fasten the training of the CNN without altering the precision and generalisation capabilities.

Fully connected layer (FC): After several layers of convolution and max-pooling, high level reasoning in the neural network is obtained via entirely connected backpropagation layers. Neurons in a fully connected layer are entirely connected to all of the previous layer outputs (as seen in regular neural networks, MLPs, Section 3.6). Their activation functions can therefore be calculated with a matrix multiplication followed by offset polarisation.

Loss layer (LOSS): The loss layer specifies how the training of the network penalises the difference between the predicted and real outputs, and is usually the last layer in the network. Various loss functions adapted to different tasks can be used there. The "Soft max" loss is used to predict one class from K mutually exclusive classes. The loss by sigmoid cross entropy is also used to predict K independent probability values in [0,1], where Euclidean loss is used to regress to actual values.

6.6.2 Choice of the hyperparameters

One of the disadvantages of CNNs, over standard MLPs, is the usage of more hyperparameters. Even if the usual learning rules rates and constants regularisation apply for both, it is necessary to take into consideration the notions of number of filters, their form as well as the form of max pooling.

Number of filters: As the size of the intermediate images decreases with the processing depth, layers near the entrance tend to have fewer filters while the layers closer to the outlet may have more. To equalise the calculation at each layer, the product of the number of characteristics and the number of pixels processed is generally chosen to be approximately nearly constant across layers. To preserve the input information, the number of intermediate outputs should be maintained (number of images multiplied by the number of pixel positions) to be increased from one layer to another. The number of intermediate images directly controls the power of the system, depending on the available examples number and the complexity of the process.

Form of the filters: Filter forms vary widely in the literature and are usually chosen based on the dataset. The best results on MNIST images (28×28) are mostly in the $5{\times}5$ range on the first layer, while the natural image datasets (often with hundreds of pixels in each dimension) tend to use larger first layer filters of 12×12 to even 15×15. The challenge is therefore to find the right level of granularity in order to create abstractions on the appropriate scale adapted to each case [10].

Form of max pooling: Typical values of max pooling are $2{\times}2$ (Figure 6.5). Very large entry volumes can justify a 4×4 pooling in the first layers. However, the choice of larger shapes will significantly reduce the signal size, and can result in the loss of too much information.

6.6.3 CNNs architectures

Convolutional neural networks can be designed and implemented in different architectures. These architectures may be classified into two categories: classical architectures and modern architectures.

Classic architectures: They are simply made up of stacked convolutional layers, such as Le-Net, AlexNet, ZFNet, and VGGNet.

LeNet [1]: The first successful applications of deep neural networks have been developed by Yann LeCun in 1998. This model was first developed to identify manuscript numbers. LeNet easily models the network layers of convolutional neurons and contains 60,000 parameters.

AlexNet [4]: AlexNet was developed and published by Alex Krizhevsky. This architecture won by a wide margin the recognition competition large-scale visual on ImageNet. The general architecture of AlexNet is similar to that of LeNet, although this model was deeper and bigger. AlexNet contains five convolution layers, three pooling layers, and two fully connected layers with 60 million settings.

ZFNet [17]: The winner of the 2014 edition of ILSV RC challenge was a convolutional network developed by Matthew Zeiler and Rob Fergus named ZFNet. This architecture is an improvement on that of AlexNet. More precisely, the modifications affected the hyper parameters of the architecture by expanding the size of the convolutional layers and reducing the size of the core on the first layer.

VGGNet [18]: The VGG network, introduced in 2014 offers a simpler but deeper variant of convolutional structures Its objective was to show that the depth of the network (number of layers) is an essential parameter to obtain good performance. It contains 138 million parameters.

Modern architectures: They explore new innovative ways to build convolutional layers in order to allow more effective learning. Among these architectures, one can cite:

GoogleNet [19]: The winner of ILSVRC was a convolutional network developed by Szegedy and his team at Google. Its peculiarity lies in the fact that it allows the application of multiple filters of different sizes on the same level. Overall, GoogleNet has three convolutional layers: two layers and nine "inception" modules [21].

ResNet [20]: Residual Networks, were developed by Kaiming He, and his team in 2015 and won the first prize in the ILSVRC competition. ResNet introduces residual blocks in which the intermediate layers of a block learn a residual function with reference to the input block. This function is a refinement step in which the characteristics map will be adjusted to obtain higher quality characteristics.

DenseNet [22]: Huang and his team published in 2016 the dense architectures. This kind of architecture has resulted in significant improvements compared to previous ones. The dense convolutional network (DenseNet) allows to connect each layer to all the others advanced layers.

6.7 Convolutional Stacked Auto-Encoder Neural Networks (CSAE)

Convolutional stacked auto-encoders, were born from the fact that fully connected AEs and DAEs both ignore a 2D image structure, while the CNN is set around the 2D structure. The problem occurs when dealing with realistically sized inputs, introducing parameter redundancy, globalising thus most features (i. e., spanning the entire visual field).

As explained for CNNs, the discovery of localised repeated features found in input data is the goal of most successful deep models. CAEs differ from conventional AEs when it comes to weight nature. In this case, weights are shared between all locations in the input, preserving spatial locality; see [23].

The reconstruction is therefore ensured by a linear combination of basic image patches based on latent code.

The CAE architecture is intuitively similar to the one described in Sections 6.5.2 and 6.5.3, except that the weights are shared. If a mono-channel input x is considered the latent representation of the kth feature map is given by

$$h^k = \sigma(x * W^k + b^k) \tag{6.25}$$

The bias b here is broadcasted to the whole map, with σ being the activation function (note that the most used functions are still sigmoidal and ReLu functions), while the operator $*$ denotes the 2D convolution. As each filter is dedicated to features belonging to the entire input, a single bias per latent map is used, as one bias per pixel would introduce a higher number of freedom degrees. The reconstruction procedure is obtained using equation (6.26):

$$y = \sigma\Big(\sum_{k\in H} \widetilde{W}^k + c\Big) \tag{6.26}$$

In this case also, there is one bias c per input channel. H represents the group of latent feature maps and $\widetilde{W}$ the flip operation over both dimensions of the weights. The 2D convolution in equation (6.25) and (6.26) is determined by context. The convolution of an $m \times m$ matrix with an $n \times n$ matrix may in fact result in an $(m + n - 1) \times (m + n - 1)$

matrix (full convolution) or in an $(m - n + 1) \times (m - n + 1)$ (valid convolution). The cost function to be minimised is the mean squared error (MSE).

For the fully connected part of the convolutional auto-encoder network, the back-propagation algorithm is applied to compute the gradient of the squared error in order to update weights as explained in Section 3.6, Chapter 3.

Similar to CNNs, CAEs use a max-pooling layer introducing sparsity over the hidden representation by erasing all nonmaximal values in nonoverlapping sub-regions, and trivial solutions such as having only one weight "on" (identity function) may be avoided. During the reconstruction phase, such a sparse latent code permits the decoding of each pixel with a minimum number of filters, by making them act more generally than specifically.

Analogously to SAE that stacks AEs, a SCAE stack CAEs. The obtained SCAE can then, prior to a supervised training stage help initialise a CNN with identical topology.

6.8 Case study: Deep neural networks for fruits recognition and classification

The aim of this section is to apply deep neural architectures, most precisely CNNs and SDAE, described in the current chapter for the classification of fruits and vegetables based on a dataset of images containing popular fruits and vegetables.

The experimental dataset used here, is taken from a larger dataset, named Fruits-360 that can be downloaded from the addresses pointed by references cited in the section. An application of CNNs using the cited dataset can be found in [25, 24], where a new, high-quality, dataset of images containing fruits is presented along with the results of some numerical experiment obtained by a particular CNN architecture. Parts of the python code used to obtain the results is presented in the paper [24] and on the corresponding GitHub space [25].

Coming back to the general image database the set contains 90483 images of 131 fruits and vegetables (as at the date of the 21st of September 2019). Noting that the dataset is constantly updated with new images of fruits and vegetables.

The original data set was created by filming the fruits while they are being rotated by a motor, while extracting frames from a 20-second short movie, with a white background as shown in Figure 6.6, here showing a grenade. Taking multiple images from the same rotating fruit shows the multiple aspect of its texture at different rotating angles is possible due to the symmetric inherent specification of fruits.

The background and lightings were not however uniformed that pushes the authors of [24] to write an algorithm to properly extract the fruit or vegetable image excluding background. The resulted dataset contains 90380 images of single fruits and vegetables spread across 131 labels. The data set is available on GitHub [25].

In this work, only 2 labels or fruit images were selected only the show the applicability and power of Deep ANNs for solving a two class classification problem. Labels

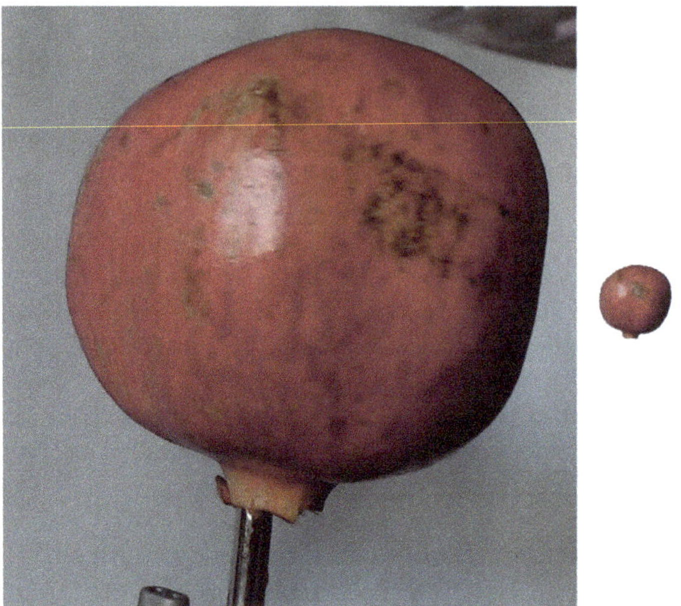

Figure 6.6: Image constitution for the fruits and vegetables database.

(a) Braeburn apple class

(b) Cherry 2 class

Figure 6.7: Two classes of fruits used: *Apple Braeburn* and *Cherry* 2.

Apple Braeburn and *Cherry*2, from the original dataset are chosen to represent respectively a type of apples and cherries; see Figure 6.7.

A total of 1640 images were used for the complete process of training, validation, and test of the classification of the two classes (Braeburn apple class, Figure 6.7(a) and cherry, Figure 6.7(b)). The dataset is divided into 492 Apple Braeburn images for the training set with 10 % used for validation and 164 images for test. The second class contains 738 images for training and validation and 246 images of Cherry 2.

The choice of the two classes, representing different fruits, is not fortuitous. Indeed, the selected apple and cherry sorts are visually close, and represents a challenge for a classification vision system, especially that the size cannot be an input characteristic, in the absence of fruit size information.

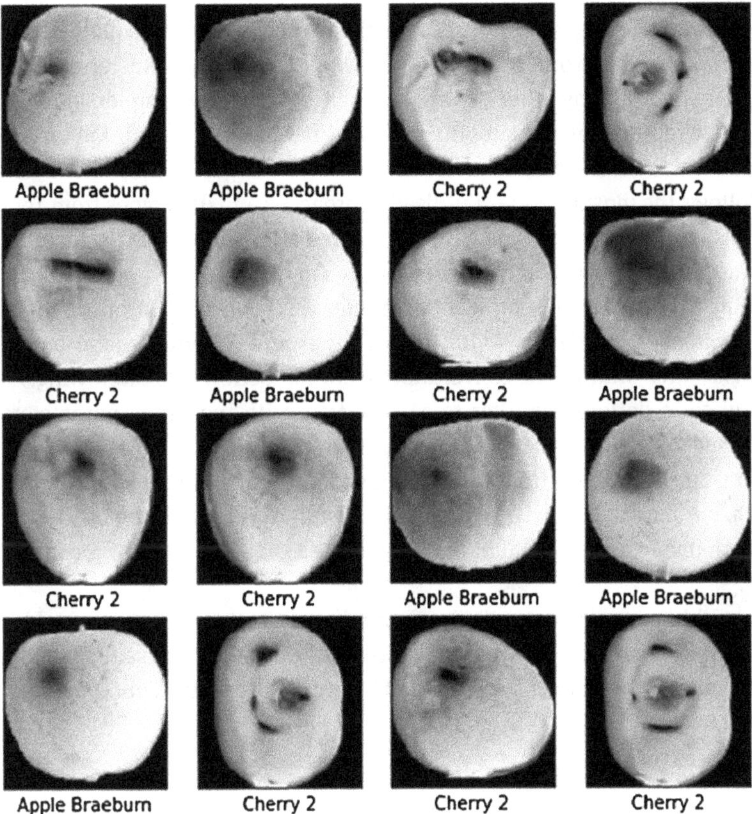

Figure 6.8: Sample images of the two used classes in grayscaling.

Figure 6.8, shows sample images for both classes in grayscaling, and shows the visual similarities between the two classes. We can see 16 sample images, hard to classify even for a human eye. The system however will be asked to provide accurate classification.

6.8.1 Fruit classification and recognition using convolutional neural networks

For the purpose of implementing, training, and testing deep neural networks, or at least some of them, the TensorFlow library is described in this section [26]. The library is available in an open source framework developed for machine learning and created by Google using data flow graphs. The graph odes represent mathematical operations, while graph edges represent the multidimensional data arrays called tensors.

The TensorFlow library offers many powerful features such as:

- allowing computation mapping for multiple machines
- it has built-in support for automatic gradient computation, etc.

For further details and description of the library, the reader is directed to [24].

For the purpose of this work, the Keras framework [27] included in TensorFlow is used. Keras provides, among others, wrappers over the operations implemented in TensorFlow. The wrappers greatly simplify calls and reduces the overall amount of code required to train and test a constructed model.

In what follows the most important utilised approaches and data types from TensorFlow for constructing and training CNNs and stacked auto-encoder CNNs, together with a short description for each of them are given, as in [24].

Convolution layer: The convolution layer computes a 2D convolution of the input of shape [*batch*, *in_height*, *in_width*, *in_channels*] and a kernel shape tensor [*filter_height*, *filter_width*]. The convolution operation performs the following:
- The function flattens the filter to a 2-D matrix with the shape parameters [*filter height* $*$ *filter width in_channels*, *output_channels*].
- Extracts image patches from the input tensor to form a virtual shape tensor [*batch*, *out_height*, *out_width*, *filter_height filter_width* $*$ *in_channels*].
- For each image patch, the filter matrix and the image patch vector are multiplied.
- If the padding is set to "same", the output keeps the same height and width as the input is said to be 0-padded. However, if the padding is set to "valid", the output may be smaller (in width and height) as the input is not 0-padded.

```
Conv2D (
      no_filters ,
      filter_size ,
      strides ,
      padding ,
      name=None
      )
```

Max pooling function: The function performs the max-pooling operation, taking as arguments the input, the filter size represents the window size, over which the max pooling function is applied and strides representing the sliding window stride for each dimension of the input tensor. Similar to the Conv2D layer, the padding parameter can set to: "valid" or "same".

```
MaxPooling2D(
      filter_size ,
      strides ,
      padding ,
      name=None
      )
```

Activation function: The function computes the specified activation function given by the operation. For this application, the rectified linear function (ReLu) is used.

```
Activation(
      operation ,
```

```
        name=None
    )
```

Dropout function: The dropout function randomly sets input values to 0 with a chosen probability *prob*.

```
Dropout(
        prob ,
        name=None
    )
```

For this application, different CNNs and SAE_CNN architectures are used to perform the classification task for fruits recognition. As previously described, Section 6.6, this type of network is constituted of convolutional layers, pooling layers, ReLU layers, fully connected layers, and loss layers.

A Rectified Linear Unit (ReLu) usually follows each convolutional layer in a typical CNN architecture. The ReLu unit is then followed by a pooling layer then one or more other convolutional layers until finally stacking one or more fully connected layers(s).

The network inputs consist of standard RGB and greyscale images of size 100×100 pixels. In what follows, results obtained from different architectures and datasets are presented.

CNN trained with Grayscale images (GrayscaleCNN): After testing a number of architectures, the best performance is obtained with CNN architecture presented in Table 6.1. Three 2D convolutional layers, one maxpooling layer, a dropout layer and a back-propagation network composed of a single hidden layer of 100 neurons.

Table 6.1: Performance of the selected Grayscale_CNN architecture.

Input	Layers	Filter	Activation function	Batch size	epochs	Loss function	optimiser
(100,100,1)	Conv2D	16*(3 × 3)	relu	32	10	Binar_cross entropy	adam
	Conv2D	16*(3 × 3)	relu				
	Conv2D	8*(3 × 3)	relu				
	MaxPooling2D	(3 × 3)	–				
	Dense	100	relu				
	Dense	1	sigmoid				

After training the performances of the Grayscale CNN are shown in Figure 6.7(a). Where the loss and accuracy for both training and validation are shown to stabilise and reach very good values after 2 epochs of training. The speed of convergence and good results is justify by the low number of classes to classify, in this case just two classes as a proof exercise of the power of CNNs.

CNN trained with RGB images: In the same manner, and after testing a number of CNN architectures using RGB images, the best performance is obtained with CNN architecture presented in Table 6.2. Three 2D convolutional layers, one max-pooling layer, a dropout layer, and a back-propagation network composed of a single layer composed of 64 neurons.

Table 6.2: Performance of the selected RGB_CNN architecture.

Input	Layers	Filter	Activation function	Batch size	epochs	Loss function	optimiser
(100,100,1)	Conv2D	16*(3 × 3)	relu	32	10	Binar_cross entropy	adam
	Conv2D	16*(3 × 3)	relu				
	Conv2D	8*(3 × 3)	relu				
	MaxPooling2D	(3 × 3)	–				
	Dense	64	relu				
	Dense	1	sigmoid				

After training the performances of the RGB_CNN are shown in Figure 6.7(b), where the loss and accuracy for both training and validation are shown to stabilise and reach very good values after 1 epochs of training. The speed of convergence and good results is justify, similar to the convergence using greyscale images, by the low number of classes to classify (2 classes).

CNN trained with grayscale and RGB images: In the same manner, and after testing a number of CNN architectures using both grayscale and RGB images, the best performance is obtained with CNN architecture presented in Table 6.3. Three 2D convolutional layers, one maxpooling layer, a dropout layer, and a back-propagation network composed of two hidden layer of respectively 128 and 20 neurons each.

Table 6.3: Performance of the selected Greyscale_RGB_CNN architecture.

Input	Layers	Filter	Activation function	Batch size	epochs	Loss function	optimiser
(100,100,1)	Conv2D	16*(3 × 3)	relu	32	10	Binar_cross entropy	adam
	Conv2D	16*(3 × 3)	relu				
	Conv2D	8*(3 × 3)	relu				
	MaxPooling2D	(3 × 3)	–				
	Dense	128	relu				
	Dense	20	relu				
	Dense	1	sigmoid				

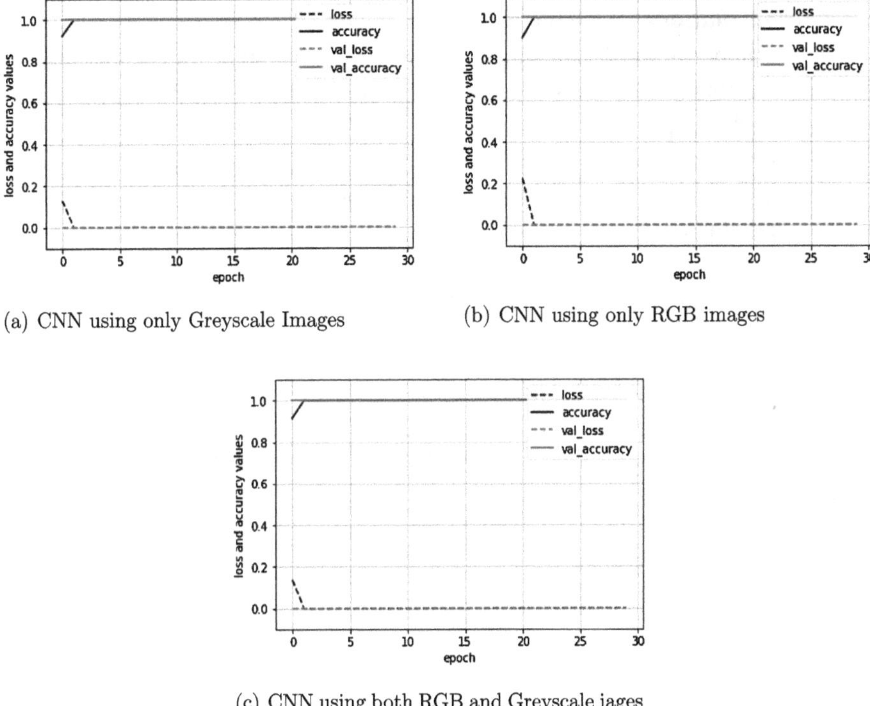

(a) CNN using only Greyscale Images

(b) CNN using only RGB images

(c) CNN using both RGB and Greyscale iages

Figure 6.9: Fruits classification loss and accuracy using CNNs.

After training, the performances of the Grey_RGB_CNN are shown in Figure 6.9(c), where the loss and accuracy for both training and validation are shown to stabilise and reach very good values after 1 epochs of training. The speed of convergence and good results is justified, similar to the convergence using grayscale images, by the low number of classes to classify (2 classes), and the obtained results are an improvement over using each type of images separately.

6.8.2 Fruit classification and recognition using stacked convolutional auto-encoder

Following the same approach, and in order to test SCAE structures, a battery of test and many architectures topologies are designed and implemented to solve the classification problem.

As previously described, Section 6.7, this type of network is constituted of stacked CAEs constituted of convolutional layers, with maxpooling additional layers and a fully connected backpropagation network. In what follows results obtained from different architectures and datasets, similar to the ones used for training CNNs are presented for the sake of objective comparison.

Stacked convolutional auto-encoders trained with Grayscale images: After testing a number of SCAEs architectures using grayscale images, the best performance is obtained stacking together two CAEs architectures followed by a fully connected backpropagation network. The complete architecture is presented in Table 6.4. Both CAEs contains four 2D convolutional layers, two of them being the transposed version of the others, one maxpooling layer and one upsampling layer. The difference of constructed CAEs are the inputs that are of reduced dimension for the second CAE, as it is obtained from the first one. The back-propagation network is finally composed of a single hidden layer of 100 neurons.

Table 6.4: Performance of the selected Grayscale_CAE architecture.

Network	Input	Layers	Filter	Activation	Batch function	epochs size	Loss function	optimiser
First CAE	(100,100,1)	Conv2D	16*(3×3)	relu	256	10	MSE entropy	adam
		Conv2D	8*(3×3)	relu				
		MaxPooling2D	(2×2)	–				
		Upsampling	(2×2)	–				
		Conv2D^T	8*(3×3)	relu				
		Conv2D^T	1*(3×3)	sigmoid				
Second CAE	(48,48,8)	Conv2D	16*(3×3)	relu	256	10	MSE entropy	adam
		Conv2D	8*(3×3)	relu				
		MaxPooling2D	(2×2)	–				
		Upsampling	(2×2)	–				
		Conv2D^T	8*(3×3)	relu				
		Conv2D^T	8*(3×3)	sigmoid				
CAE	(22,22,8)	dense	100	relu	32	10	Binary Rosentropy	adam
		dense	1	sigmoid				

After training, the performances of the Gray_SCAE are shown in Figure 6.10(a), where the loss and accuracy for both training and validation are shown to stabilise and reach very good values after 7 or 8 epochs of training. The complexity does not seem to justify the obtained results over CNN architectures.

Stacked convolutional auto-encoders trained with RGB images: The same procedure is repeated with RGB images, the best performance is obtained stacking together two CAEs architectures followed by a fully connected back-propagation network. The complete architecture is totally similar to the one used with Grayscale images, presented in Table 6.4. The performances, however, are greatly improved with a convergence after only one epoch of training, as shown in Figure 6.10(b).

Stacked convolutional auto-encoders trained with Grayscale and RGB images: The same procedure is repeated with RGB and Grayscale images, the best perfor-

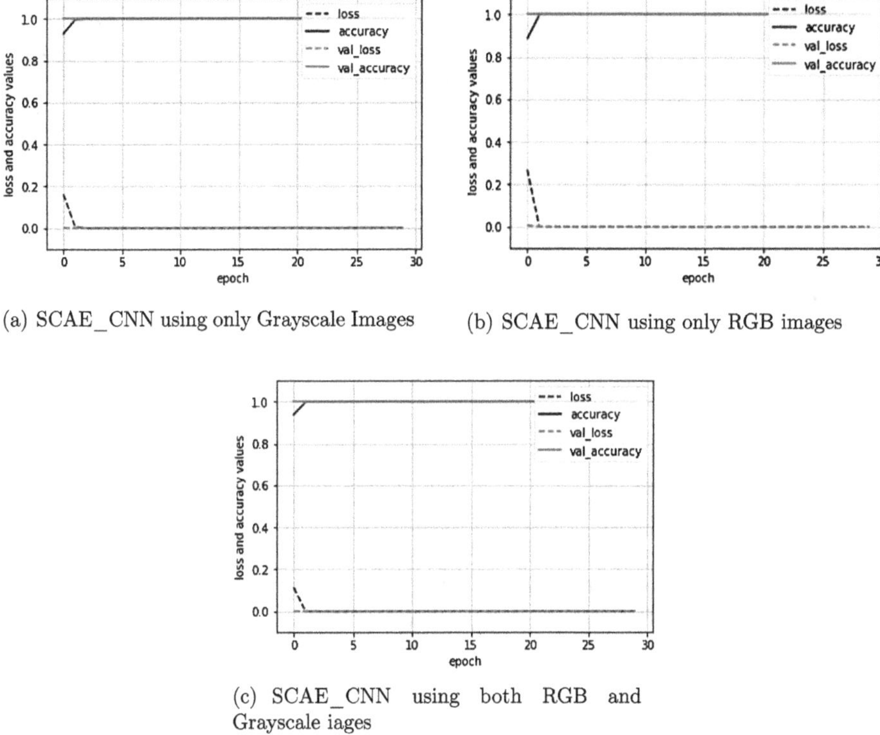

(a) SCAE_CNN using only Grayscale Images

(b) SCAE_CNN using only RGB images

(c) SCAE_CNN using both RGB and Grayscale iages

Figure 6.10: Fruits classification loss and accuracy using stacked auto-encoder CNNs.

mance is obtained stacking together two CAEs architectures followed by a fully connected back-propagation similar to the one used with Grayscale images, presented in Table 6.4. The performances obtained are very satisfactory after only one epoch of training, as shown in Figure 6.10(c).

6.8.3 Fruit classification and recognition using stacked convolutional denoising auto-encoder

In this section, the same SCAE architectures are trained using additional noise in order to form Stacked Convolutional Denoising Auto-Encoders (SCDAE). Introducing white noise as for SDAE (see Section 6.5.3) favors training the network on data on altered inputs conferring it, hopefully, a better generalisation. The obtained results showed an improvement of 0.01 % over SDAE architectures, not justifying the added training complexity.

Table 6.5 summaries all deep architecture results using either Grayscale and RGB images alone or both datasets. It can be seen that the results are very comparable and

Table 6.5: Recognition percentage and classification records for different deep network architectures.

Network	Image	Accuracy	Apple Class		Cherry Class	
			Well classified	Misclassified	Well classified	Misclassified
	Grayscale	0.921	164	0	214	32
CNN	RGB	0.924	164	0	215	31
	RGB and Grayscale	0.929	164	0	217	29
	Grayscale	0.953	164	0	227	19
SCAE	RGB	0.748	164	0	143	103
	RGB and Grayscale	.917	164	0	212	34
	Grayscale	0.953	162	2	227	19
SCDAE	RGB	0.951	164	0	226	20
	RGB and Grayscale	0.953	164	0	227	19

the improvements of an architecture over another, for the classification of two classes only, are minimal. However, the best performance is achieved by the CNN using both Grayscale and RGB images. This can be justified by the inherent ability of CNNs to deal with natural 2D data, with particular images.

Bibliography

[1] Le Cun, Y., Bottou, L., Bengio, Y., Haffner, P., et al. (1998). Gradient-based learning applied to document recognition. Proceedings of the IEEE, 86(11), 2278–2324.

[2] LeCun, Y., Bengio, Y., and Hinton, G. (2015) Deep learning. Nature 521, 436–444.

[3] Meyer, P. (2018) Vers la Modèlisation des traitements de radiothèrapie: Apprentissage profond en radiothèrapie. Mémoire de HDR, Ecole Doctorale de Physique, Université de Strasbourg, France.

[4] Krizhevsky, A., Sutskever, I., and Hinton, G. E. (2012) Imagenet classification with deep convolutional neural networks, in Advances in Neural Information Processing Systems, pp. 1097–1105.

[5] Dodge, S. and Karam, L. (2017) A study and comparison of human and deep learning recognition performance under visual distortions, in 2017, 26th International Conference on Computer Communication and Networks (ICCCN), pp. 1–7.

[6] Bayle, Y. (2018) Apprentissage automatique de caractéristiques audio: application à la gènèration de listes de lecture thèmatiques. PhD thesis.

[7] Litjens, G., Kooi, T., Bejnordi, B. E., Setio, A. A., Ciompi, F., Ghafoorian, M., Van Der Laak, J. A., Van Ginneken, B., and Sánchez, C. I. (2017) A survey on deep learning in medical image analysis. Medical Image Analysis.

[8] Hochreiter, S. and Schmidhuber, J. (1997) Long short-term memory. Neural Computation, 9(8), 1735–1780.

[9] Hochreiter, S., Bengio, Y., and Frasconi, P. (2001) Gradient flow in recurrent nets: the difficulty of learning long-term dependencies, in J. Kolen and S. Kremer, Eds., Field Guide to Dynamical Recurrent Networks. IEEE Press.

[10] Deng, L. (2012) The mnist database of handwritten digit images for machine learning research [best of the web]. IEEE Signal Processing Magazine, 29(6), 141–142.
[11] Gers, F. A., Schmidhuber, J. A., and Cummins, F. A. (2000) Learning to forget: continual prediction with LSTM. Neural Computation, 12(10), 2451–2471.
[12] Gers, F. A., Schraudolph, N. N., and Schmidhuber, J. (2002) Learning precise timing with LSTM recurrent networks. Journal of Machine Learning Research, 3(Aug), 115–143.
[13] Vincent, P., Larochelle, H., Lajoie, I., Bengio, Y., and Manzagol, P. A. (2010) Stacked denoising autoencoders: learning useful representations in a deep network with a local denoising criterion, Journal of Machine Learning Research, 11, 3371–3408.
[14] Hinton, G. E., Krizhevsky, A., and Wang, S. D. (2011) Transforming auto-encoders, in International Conference on Artificial Neural Networks, Springer, pp. 44–51.
[15] Larochelle, H., Erhan, D., Courville, A., et al. (2007) An empirical evaluation of deep architectures on problems with many factors of variation, in Proceedings of the 24th International Conference on Machine Learning, ACM, pp. 473–480.
[16] Graham, B. (2014) Fractional max-pooling. arXiv preprint arXiv:1412.6071.
[17] Zeiler, M. D. and Fergus, R. (2014) Visualizing and understanding convolutional networks, in EurOpean Conference on Computer Vision, Springer, pp. 818–833.
[18] Simonyan, K. and Zisserman, A. (2014) Very deep convolutional networks for large-scale image recognition. arXiv preprint arXiv:1409.1556.
[19] Szegedy, C., Vanhoucke, V., Ioffe, S., Shlens, J., and Wojna, Z. (2016) Rethinking the inception architecture for computer vision. In Proceedings of the IEEE Conference on Computer Vision and Pattern Recognition, pp. 2818–2826.
[20] He, K., Zhang, X., Ren, S., and Sun, J. (2016) Deep residual learning for image recognition. In Proceedings of the IEEE Conference on Computer Vision and Pattern Recognition, pp. 770–778.
[21] Shin, H. C., Roth, H. R., Gao, M., Lu, L., Xu, Z., Nogues, I., Yao, J., Mollura, D., and Summers, R. M. (2016) Deep convolutional neural networks for computer-aided detection: CNN architectures, dataset characteristics and transfer learning. IEEE Transactions on Medical Imaging, 35(5), 1285–1298.
[22] Huang, G., Chen, D., Li, T., Wu, F., Van Der Maaten, L., and Weinberger, K. Q. (2017) Multi-scale dense convolutional networks for efficient prediction. arXiv, 35(2).
[23] Masci, J., Meier, U., Ciresan, D., and Schmidhuber, J. (2011) Stacked convolutional auto-encoders for hierarchical feature extraction, in ICANN 2011, Part I, LNCS 6791, pp. 52–59.
[24] Muresan, H. and Oltean, M. (2018) Fruit recognition from images using deep learning, Acta Univ. Sapientiae, Informatica, 10(1), 26–42.
[25] Oltean, M., and Muresan, H. Fruits 360 dataset on github. [Online; accessed 19.05.2020]. 1, 10,.
[26] Abadi, M., et al. (2015) TensorFlow: Large-scale machine learning on heterogeneous systems. Software available from tensorflow.org.
[27] Gulli, A. and Pal, S. (2017) Deep learning with Keras, Packt Publishing Ltd.

7 Overview of model predictive control theory and applications in food science using ANN

This chapter focuses on Model Predictive Control (MPC), more precisely MPC using Artificial Neural Networks and will focus on food technology and food processing applications. The principles of MPC will therefore be stated and explained followed by the derivation of the corresponding control laws. The chapter will then present an overview of the most reported MPC applications using ANNs and will depict a real life example taking from food industry.

7.1 ANN and model predictive control in food process applications

Section 2.5 presented a thorough overview of ANN applications in the food industry. The present section will focus, however, on a more particular aspect of ANNs usage in process control strategies, more precisely on the usage of ANNs with the MPC paradigm for food processes.

MPC is a mathematically formulated scheme, used to predict and control the process behaviour allowing flexible and efficient exploitation of the target and output signal understanding. This helps produce optimal performances of a system under various constraints, just like an optimal control with a prediction receding horizon. This led to a wide number of applications in agriculture and food processing [6, 5].

The need to handle some difficult control problems, led the control community to investigate machine learning capabilities and use them in control strategies. The use of ANNs in MPC has since the 1990s attracted a great deal of attention and the trend is still rising. The efficacy of neural predictive control lies in the fact that the performed predictions are obtained by a nonlinear model, that is, in general closer to the behaviour of the real nonlinear process when compared to a simplified linear model. The Neural Network Model Predictive Control (NNMPC) method is less sensitive to perturbations and oscillations, principally due to nonlinear process behaviour and when dealing with noise when compared to PID and PI controllers as well as linear MPC.

Indeed, in practice, control engineers almost always resorted to linear approximations. This is achieved via simplified mathematical modelling or data based approaches such as: pseudo-random process excitation and step test in order to identify an input/output model that best approximates the process behaviour. The obtained model has then to be validated using separate plant data.

The resulting models are often adapted for the design of simple classical controllers (such as PID controllers), and suit fixed set points around a linear behaviour range of the process. The model, however, is enable to accurately approximate a highly nonlinear process. Often, an approximate first-principles nonlinear model exists, but

https://doi.org/10.1515/9783110646054-007

is, in most cases, too effort and time consuming as well as nonadaptable to implement classical model based control approaches.

One of the alternatives to predict the behaviour of complex nonlinear processes is provided by ANN models based on their historical data. As detailed in Section 2.5, and unlike other models, neural networks are capable of generalisation and providing accurate predictions even if all mechanisms and principles affecting the behaviour of the processes are not understood. Moreover, ANNs have the ability of inferring general rules as well as extracting typical pattern from specific examples as well as recognising input-output mapping parameters from complex multi-dimensional data. The interpolation between typical patterns as well as the learning generalisation to nearby patterns and phenomenon permitting knowledge extrapolation beyond training data, make ANNs a better candidate for approximating highly nonlinear processes. Reminding the reader that food processes control consist, in most cases, in flow and temperature control and addresses highly nonlinear problems such as, thermodynamic, heat transfer, chemical reactions, etc.

7.1.1 Strategy of using ANNs in a MPC approach

Many surveys have been carried out on how to use and incorporates ANNs into control approaches and are given in [8]. MPC is one of the most suitable control algorithm permitting the usage of ANN models as an internal predictive model. The idea of incorporating neural network models in MPC algorithms has been proposed by a number of researchers. In what follows, the most successful and reported approaches are cited:

Bhat and McAvoy (1990) [9]: The linear model is replaced by a dynamical ANN model. The authors give some advice on how to choose the nonlinear optimisation routine needed to estimate the optimal control action.

Lee and Park (1992) [10]: Here, the ANN is used to learn the disturbance dynamics, and was tested on a simulated distillation column and reactor control.

Willis et al. (1991) [11]: Modelled a distillation column using ANN, and used the model as an internal MPC model.

Declerq et al. (1996) [12]: Here, the authors used three different ANN models namely the: feedforward, radial, and Elman forms. These ANNs were then used in model based predictive control algorithms. Model time-validation approach is used. The feedforward ANN is found to better estimates the inherent nonlinearity of the system sensibly better than the other ANN topologies (namely Elman and RBF networks).

Hernandez and Arkun (1990) [13]: developed an extended version of the Dynamic Matrix Control (DMC) approach, able to stabilise a system around an open-loop unstable point, where the linear DMC version was not successful when performing the same task.

The reader can find a comprehensive review of MPC using ANN and their applications in chemical processes, classified in: Batch reactors, distillation columns, bio-process systems and others [7]. The author will try to give the most complete review of MPC using ANNs in the field of food engineering and food processing in the following section.

7.1.2 Overview for ANN based MPC in food processes

For food and bioprocesses, physical modeling is found to be a complex, difficult, and time consuming procedure. As said earlier, this encouraged the use of data driven modelling in general and ANNs in particular. A comprehensive review for MPC, including ANN based MPC can be found in [14]. In what follows, part of this review will be related as well as other references and works.

7.1.3 Baking applications

Table 7.1 shows the applications reported in baking using ANN based MPC. The author reported two and one applications on food extrusion processes and bread baking process respectively. In [30], the startup phase of a food extrusion process is modelled by a neural network model, where Paquet-Durand et al. in 2012 [15] uses digital image processing for controlling and monitoring a baking process. A neural network was then used to identify the state of the baking process, based on the lightness and colour saturation of the baked products during baking.

Table 7.1: List of ANN based MPC in baking processes.

Baking Applications	Control Approach	References
Bread baking process	Neural network-based control (digital image processing)	[15]
Food extrusion process	ANN based control	[16]
	ANN and expert system	[30]

The last reported application in baking is given by [16]; here, a number of ANNs with output feedback were used for the first time for the control of Specific Mechanical Energy (SME). The control variable is given by the screw speed in flat bread production in a twin-screw extrusion cooker.

7.1.4 Drying applications

As for baking applications, there are three reported applications in drying. Yüzgeç et al., in 2008 [17] used a dynamic neural network based model predictive control for a baker's yeast drying process. The optimal drying profile for the drying process are determined using a genetic algorithm. This control structure resulted in a better performance compared to a nonlinear partial differential equations model and a feedforward neural network only. The drying applications of food processes using ANN MPC based controller are presented in Table 7.2.

Table 7.2: List of ANN based MPC in drying processes.

Drying Applications	Control Approach	References
Baker's yeast drying process	Dynamic neural network-based model	[17]
	Adaptive neuro-fuzzy inference system (ANFIS)	[18]
Fish meal drying process	Artificial neural network based control	[19]

Adaptive Neuro-Fuzzy Inference System (ANFIS) has been used in the control of industrial-scale batch drying of baker's yeast in [18], and based on a hybrid machine learning algorithm using one-year industrial scale data under various working constraints and settings.

The last in date reported application, is given in [19]. For the matter, a combination of first principle mathematical and ANNs is used for parameter estimation for a fishmeal rotary drying process. A model predictive controller, adjusting the inlet air temperature to control the moisture content of fishmeal product is developed.

The simplest MPC control structure given by a fixed extended horizon model predictive controller was successfully employed to an extended range of the controlled variables. Note that the quadratically constraints, quadratic dynamic matrix control (QDMC) algorithm failed to give satisfactory results.

7.1.5 Fermentation/brewing applications

Eight applications of Neural MPC controllers are reported in fermentation and brewing processes in different aspects and for different products. The reported applications of ANN based MPC in fermentation and brewing are presented in Table 7.3.

In complex fermentation and brewing process control applications, ANN are the perfect candidates of taking advantage of the data logged from online sensors, real-time state estimation and predictions [22–25]. Based on industrial-scale fedbatch fermentation data. A neural network was used for the estimation and the control of consumed sugar and total lysine production amount [26].

Table 7.3: List of ANN based MPC in fermentation/brewing processes.

Fermentation/Brewing Applications	Control Approach	References
Lactic acid production	Neural network-based model-based control (nonlinear model predictive control—NMPC)	[34]
Control of yeast fermentation for bio-reactors	Neural network-based control	[25]
	Feedforward neural network-based control	[21]
Total lysine production	Neural network-based control	[26]
Yogurt production	Neural network-based control	[32]
Production of lipase enzyme by Candida rugosa	Neural network-based control	[29]
Control of fed-batch Δ-galactocidase fermentation	Neural network-based control	[28]
Control of alcoholic fermentation process	Neural network-based control model-based control (MPC)	[40]

In [27], an ANN for the fermentation estimation time in the production of β-galactocidase by Streptococcus *salivarius* subsp *thermophilus* 11F is presented. Controlling the fed-batch β-galactocidase fermentation by recombinant Escherichia coli due to its high sensitivity to influx, noise, and process burdens is given in [28]. Beer brewing [31], yogurt production [32], and enzyme production [29] are other reported ANN control applications. Also, a real time as well as neuro-fuzzy state approximation, prediction and control for baker's yeast fermentation is presented in [33].

More recently, [34] showed the design and analysis of a nonlinear model predictive control (NMPC) approach for a lactic acid production. The control was conducted linking two stirred bio-reactors sequentially.

7.1.6 Thermal/pressure applications

Applications of advanced process control systems to thermal and high-pressure food processing are of paramount importance as most processes in food industry, and are based on temperature and/or pressure systems. [20] investigates ANN capabilities in estimating process variables in thermal/high pressure food processing.

The five reported application of ANN based MPC are listed in Table 7.4, and are described in what follows.

O′Farrell et al., in 2004, worked on food colour recognition (chicken wings, whole chicken, cereal patti burgers, sausages, etc.). The authors presented an ANN based of fiber optic spectroscopic technique obtained data in a large scale industrial oven [35].

Table 7.4: List of ANN based MPC in thermal and/or pressure processes.

Thermal/pressure Applications	Control Approach	References
Thermal sterilization of solid canned foods	Neuro-fuzzy MPC	[41]
Cane sugar/juice production	MPC with ANN model	[36]
	Intelligent ANN controller	[37]
Control in a large-scale industrial oven	ANN based Control	[35, 42]
Tubular heat exchanger for Thermal food processing	ANN based Control	[38]
Thermal processing of ginseng	ANN based Control	[39]

For the application, an ANN is developed to predict cooking colour of the product and help control the optimal cooking, monitoring the predicted colour.

Benne et al., (2000) [36], developed an ANN based MPC to control the evaporation process of sugar cane. A MPC technique based on a multi-step predictor through a neural network model of the plant, is developed in order to achieve more robust control. For the same application, a control scheme for a multiple effect evaporator in order to increase the concentration (brix) of sugar cane juice from a nominal value of 15 wt% to syrup with a brix of 72 wt% is presented in [37].

Vasièkaninovà et al., (2011) [38] developed MPC controllers using ANN based internal models for the control of temperature in tubular heat exchangers.

Martynenko and Yang in 2007 [39], proposed an intelligent control system based on ANNs for thermal processing of natural bio-materials based on different sensor fusions and machine vision.

7.1.7 Dairy applications

MPC has found wide usage in dairy industry, as for industrial processes in general as cited in [47]. Most applications of MPC in dairy up to 2008 are reported in [1], without highlighting ANN base MPC.

Four applications depicting the usage of ANN based MPC in dairy processes are given in Table 7.5. Two of them concerns milk pasteurisation in plate heat exchangers and some of the obtained results will be presented by the author in Section 7.5. Indeed, Khadir and Ringwood, 2003 [3, 4], in order to increase the performance of linear model predictive control and classical control performances for milk pasteurisation in Plate Heat Exchangers (PHE), developed an ANN based MPC controller and tested it in the presence of disturbances, set point changes, and parametric uncertainties.

Table 7.5: List of ANN based MPC in thermal dairy processes.

Dairy Applications	Control Approach	References
Milk pasteurisation process	ANN model Based MPC	[3, 4]
Plate heat exchanger for milk pasteurisation	Fuzzy ANN based MPC	[43]
pH and acidity for industrial cheese production	ANN based Model	[2]

For the same purpose, [43] used neural network and fuzzy logic for the control of pasteurisation temperature in a Plate Heat Exchanger (PHE). AI approaches are particularly valuable and efficient when dealing with processes exhibiting highly on linear behaviour as well as high dimensions and multiple raw materials.

In the last application, a reduced 9 inputs neural network model was able to predict the final cheese pH with the same accuracy as a complete physical model with 33 original input variables in [2].

7.1.8 Other food process applications

The rest of the reported applications are classified as miscellaneous, and just two applications are reported in Table 7.6.

Table 7.6: List of ANN based MPC in other food processes.

Other Applications	Control Approach	References
Rice cake production	Fuzzy ANN control	[44]
Coffee grinding	ANN based control	[45]

Kupongsak and Tan in 2006 [44], used fuzzy logic and ANN based on desirable sensory attributes to derive setpoint for rice cake products. Mesin et al., in 2012 [45], studied the control of coffee grinding with ANNs, where one MLP was used to control two coffee production grinders.

7.2 Principles of linear model predictive control

7.2.1 Introduction

Model Predictive Control (MPC) regroups under the nomination a class of control algorithms, based on the calculation of an "optimal" control variable for the regulation or set point tracking of, future plant behaviour over a defined prediction horizon. MPC saw its birth in the petroleum power plants and refinery [46]. The MPC technology has since spread to many others industrial fields [47]. The control engineer and practitioner is then presented with a large number of algorithms respecting MPC principles, having different control law formulation based on different internal model representations. Choosing a particular MPC approach over another, must rely on strong knowledge of the selected MPC mathematical formulation as well as an extensive insight of the process behaviour and a deep understanding of the data representing it.

Control of any industrial process can be fitted in any stage of the layered approach shown in Figure 7.1. PID and its control versions, such as ratio control, feedforward, etc, can be classified in layer 3 and 4 respectively. Dynamic control and optimization can be classified in layer 5, note that MPC may be considered as a dynamic or an optimization control approach.

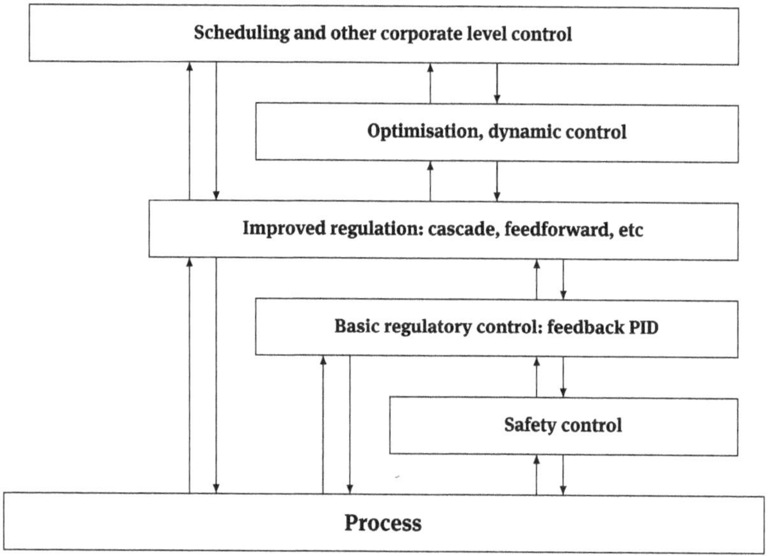

Figure 7.1: Control layered hierarchy approach.

The primary purpose of this chapter is to familiarise the reader with the concept of MPC, its origin and history, as well as describing briefly the most popular algorithms in order to justify the choice(s) for controlling food processes.

Although this chapter provides a brief overview of MPC history, more complete reviews on MPC theoretical aspects can be found in [54, 59–62] and [63].

In addition to this, Qin and Badgwell, in 2003 [47] focused on reviewing commercially available industrial MPC algorithms.

7.2.2 Origins of model predictive control

It took Jean Piaget (1976) nearly 20 years to convince the scientific community to adopt his theory on how the human brain operates, (e. g., to coordinate and achieve physical movements in order to accomplish a given task). Piaget's theory could be based on the following features:
1. creation of an operative image
2. desired future
3. action
4. assessment of the action (i. e., comparison of the operative image and the real action).

The process of achieving a task becomes then, a recursion of the four above cited steps, even for routinely daily actions. During our daily activities these steps are repeated over and over again and seem natural, despite being the result of a thinking process [48].

In 1968, Jacques Richalet established an analogy between a possible new control technique and Piaget theory. Four principles corresponding to: the operative image, desired future, action, and comparison were then established as a prelude to MPC theory. Piaget steps in human thinking were then replaced by four principles more appropriate for computer programming:
- **Internal model:** The embedded internal MPC model must be able to accurately predict the process future outputs over a defined prediction range.
  The internal model, as it has to be implemented on a calculator, has to be formulated in discrete-time. However, the internal model formulation may be of different forms and formulations. It can be linear, nonlinear, transfer function form, state space form, first principles, black-box, etc.
- **Reference trajectory:** The reference trajectory consists in a future desired output trajectory based on the measured output value. The controller will not then aim at the set-point itself, whatever constant or time varying, but will aim at the reference trajectory leading to that set-point. This concept is immediately known as the closed loop behaviour. Moreover, this reference trajectory can be time or state-varying and is open to all kinds of specific solutions differing from one MPC algorithm to another (Section 7.2.2).
- **Calculation of the manipulated variable:** Again, the means of establishing an "optimal" manipulated variable (MV) differs from one MPC algorithm to another. The main expense of MPC over classical control, the optimization of the future MV, with respect to the minimization of an error criterion in order to obtain the desired plant behaviour. The optimization of the error criterion leads to an analytical or numerical solution, depending on the nature of the internal model (resp., linear and nonlinear internal models).

- **Auto-compensation:** Despite the most accurate optimization, there always will be a difference between the model and the real process outputs. This difference, given by an error, can be used to assess the model's accuracy. This error can be used to improve the control quality. The error is generally not null, for the following reasons:
  - the physical process is generally altered by unknown parameters, which impose random changes on the output,
  - the model is always relatively inaccurate.

  In order to minimise this error, several techniques are used in MPC theory for example, state estimation and on-line adaptation of the model parameters.

7.3 Linear model predictive control algorithms

Having described the general principles of MPC, the author feels important to give the mathematical formulation of the two most popular MPC algorithms as a pedagogical effort. Indeed, they summarise the foundation and the step taking in any MPC formulation even using a nonlinear internal model. The two MPC algorithms presented are:

- Predictive Functional Control (PFC), in its simplest form.
- Generalised Predictive Control (GPC).

Note that only Single Input Single Output (SISO) forms of the algorithms are given here. The reader may refer to [51, 52] and [50] for Multi-Input Multi-Output (MIMO) PFC and GPC formulations.

7.3.1 First-order predictive functional control

Based on [49], a first-order process is considered and given by the following transfer function:

$$G_P(s) = \frac{K}{1 + \tau s} \tag{7.1}$$

If we consider the following characteristics for the MPC formulation:

- a setpoint given by a constant (regulation case): $C(k) = C_0$
- a simple step basis function
- an reference trajectory expressed by an exponential $\lambda = e^{(-\frac{T_S}{T_R})}$, where: T_S is the sampling period and T_R is the response time in closed loop to be specified, and
- a single coincidence point H.

Once the MPC design parameters are set, the subsequent steps are the design and the formulation law of the controller.

Step 1: Model formulation
If the model is given by the following transfer function:

$$G_M(s) = \frac{K_M}{1 + \tau_M s}$$

The zero-order equivalent of $G_M(s)$ leads to the discrete formulation of the model on the form of the difference equation (7.2).

$$y_M(k) = \alpha y_M(k-1) + K_M(1-\alpha)u(k-1) \tag{7.2}$$

where $\alpha = e^{(-\frac{T_S}{\tau_M})}$.

If the basic function is a step (zero-order basis function) then:

$$y_A(k+H) = \alpha^H y_M(k) \tag{7.3}$$

$$y_F(k+H) = K_M(1-\alpha^H)UB_0 = K_M(1-\alpha^H)u(k) \tag{7.4}$$

where, y_A and y_F are the free (auto-regressive) and the forced response of y_M, respectively.

Step 2: Reference trajectory formulation
If y_R expresses the reference trajectory formulation, then at the coincidence point H we expect the process output y_P to be equal to y_R:

$$C(k+H) - y_R(k+H) = \lambda^H(C(k) - y_P(k))$$

thus:

$$y_R(k+H) = C(k+H) - \lambda^H(C(k) - y_P(k)) \tag{7.5}$$

Step 3: Predicted process output
From equation (7.5), the predicted model output using auto-compensation is given as:

$$\hat{y}_P(k+H) = y_M(k+H) + (y_P(k) - y_M(k)) \tag{7.6}$$

Step 4: Computation of the control law
At the coincidence point H we require that

$$y_R(k+H) = \hat{y}_P(k+H)$$

Using equations (7.2), (7.3), (7.4), and (7.5), we obtain

$$C(k+H) - \lambda^H(C(k) - y_P(k)) - y_P(k) = y_M(k+H) - y_M(k) \tag{7.7}$$

In the regulation case, where the reference set-point is a constant, $C(k+H) = C(k)$, rearranging equation (7.7) permits to obtain

$$C(k)(1-\lambda^H) - y_P(k)(1-\lambda^H) + y_M(k)(1-\alpha^H) = K_M(1-\alpha^H)u(k) \tag{7.8}$$

When solving for $u(k)$, the final control law is then given by equation (7.9):

$$u(k) = \frac{(C(k) - y_P(k))(1-\lambda^H)}{K_M(1-\alpha^H)} + \frac{y_M(k)}{K_M} \tag{7.9}$$

The block diagram description of the obtained control law is represented by Figure 7.2.

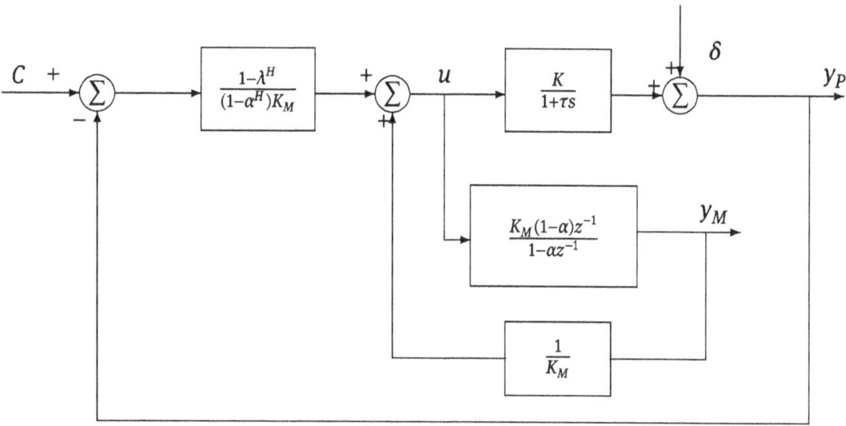

Figure 7.2: Formulation of the control law for a first-order PFC.

7.3.2 Generalised predictive control

Generalised predictive control (GPC) was developed in 1985, and was intended to offer a new control alternative. GPC is based on a transfer function (input/output model) auto-regressive and integrated moving-average (CARIMA) type model; given in equation (7.10) [53].

$$A(q^{-1})y(k) = B(q^{-1})u(k-d) + C(q^{-1})\frac{e(k)}{\Delta} \tag{7.10}$$

where d is the pure delay index, (i.e., number of samples), Δ is the delta operator $(1-q^{-1})$, $e(k)$ is the disturbance signal and $A(q^{-1})$, $B(q^{-1})$ are the system polynomials.

Where, A, B, and C are polynomials in the backward shift operator q^{-1}:

$$B(q^{-1}) = b_0 + b_1 q^{-1} + \cdots + b_{nb} q^{-nb}$$

$$A(q^{-1}) = 1 + a_1 q^{-1} + \cdots + a_{na} q^{-na}$$
$$C(q^{-1}) = 1 + c_1 q^{-1} + \cdots + c_{nc} q^{-nc}$$

$e(k)$ is an uncorrelated random sequence (disturbance signal) and Δ is the delta operator $(1 - q^{-1})$.

In the basic GPC algorithm, $C(q^{-1})$ is chosen set to 1, for simplicity. In the general case $C(q^{-1})$ is truncated and absorbed by $A(q^{-1})$ and $B(q^{-1})$ (Clarke et al. (1987)). The CARIMA model then became

$$A(q^{-1})y(k) = B(q^{-1})u(k-1) + \frac{e(k)}{\Delta(q^{-1})} \tag{7.11}$$

For the derivation of a j-step ahead predictor of $y(k+j)$ based on (7.11), consider the identity:

$$1 = E_j(q^{-1})A(q^{-1})\Delta(q^{-1}) + q^{-j}F_j(q^{-1}) \tag{7.12}$$

where E_j and F_j are polynomials uniquely defined given $A(q^{-1})$ and the prediction interval j. If equation (7.11) is multiplied by $E_j(q^{-1})\Delta(q^{-1})q^{-j}$, we have

$$E_j(q^{-1})A(q^{-1})\Delta(q^{-1})y(k+j) = E_j(q^{-1})B(q^{-1})\Delta(q^{-1})u(k+j-1) + E_j(q^{-1})e(k+j) \tag{7.13}$$

Dropping the q^{-1} dependency for simplicity, and using the identity in equation (7.12). Rewriting equation (7.13) in order to isolate $y(k+j)$, we obtain

$$y(k+j) = E_j \Delta B u(k+j-1) + F_j y(k) + E_j e(k+j) \tag{7.14}$$

As $E_j(q^{-1})$ has a degree of $j-1$, and all noise components are predicted and present in the future. The optimal predictor provided by the measured output data up to time t and any given $u(k+i)$ for $i > 1$, is given by

$$\hat{y}(k+j|t) = G_j \Delta u(k+j-1) + F_j y(k) \tag{7.15}$$

where $G_j(q^{-1}) = E_j B$.

In order to implement a long-range prediction given in equation (7.15), the identity equation (7.13) has to be resolved numerically for E_j and F_j, and this for the entire considered range of j.

An appropriate solution is, to use the recursion of the Diophantine equation in (7.12), so that the polynomials E_{j+1} and F_{j+1} are obtained, given the values of E_j and F_j.

Denoting $E = E_j$, $R = E_{j+1}$, $F = F_j$, $S = F_{j+1}$, consider two Diophantine equations with $\tilde{A} = A\Delta = 1 + \tilde{a}_1 q^{-1} + \cdots + \tilde{a}_{na} q^{-na+1}$:

$$1 = E\tilde{A} + q^{-j}F \tag{7.16}$$
$$1 = R\tilde{A} + q^{-(j+1)}S \tag{7.17}$$

Subtracting (7.16) from (7.17) gives

$$0 = \tilde{A}(R - E) + q^{-j}(q^{-1}S - F)$$

The polynomial $R - E$ is of degree j and may be split into two parts:

$$R - E = \tilde{R} + r_j q^{-1}$$

The solution is then $\tilde{R} = 0$ and also S given by $S_q(F - \tilde{A}r_j)$.
$\tilde{A}$ has a unit leading element and we have

$$r_j = f_0 \tag{7.18}$$

$$s_i = f_{i+1} - \tilde{a}_{i+1} r_j \tag{7.19}$$

for $i = 0$ to the degree of $S(q^{-1})$; and

$$R(q^{-1}) = E(q^{-1}) + q^{-1} r_j \tag{7.20}$$

$$G_{j+1} = B(q^{-1})R(q^{-1}) \tag{7.21}$$

Considering the process polynomials $A(q^{-1})$ and $B(q^{-1})$ and the solution $E_j(q^{-1})$ and $F_j(q^{-1})$, equations (7.18) and (7.19) can be used to obtain $F_{j+1}(q^{-1})$, and (7.20) to give $E_{j+1}(q^{-1})$ and so on. To initialise the iterations note that for $j = 1$:

$$1 = E_1 \tilde{A} + q^{-1} F_1$$

and as the leading element of $\tilde{A}$ is 1 then:

$$E_1 = 1, \quad F_1 = q(1 - \tilde{A})$$

Assuming that a future reference sequence, based on an original set-point $[w(k + j); j = 1, 2, \ldots N_h]$ is available, with in most cases $w(k + j)$ being a constant w, i. e., regulation case.

A future plant outputs $y(k + j)$ as close a possible to w, is the objective of the computed future control law.

A receding-horizon approach is then used where at each simple-instant, k the following actions are realised:

1. The future reference trajectory sequence is calculated based on a given set-point.
2. A set of predicted outputs $\hat{y}(k + j|j)$ with corresponding predicted system error $e(k + j) = w(k + j) - \hat{y}(k + j|j)$ is generated by the prediction model in equation (7.15). Note that $\hat{y}(k + j|j)$ for $j > k$ depends in part on future control signals $u(k + i)$ that are to be determined.
3. A criterion containing the future computed errors is minimised over a “control horizon”, where further $u(k + i)$ are not required.

4. A sequence of future control actions is then calculated and the first element $u(k)$ of the sequence is selected. Consequently, the appropriate data vectors is shifted so that the calculations can be repeated at the next sample instant.

If the following cost function is considered:

$$J(N_1, N_2) = \sum_{j=N_1}^{N_2} [w(k+j) - y(k+j)]^2 + \sum_{j=1}^{N_2} \lambda(j)[\Delta u(k+j-1)]^2 \tag{7.22}$$

where:
N_1: representing the minimum costing horizon.
N_2: representing the maximum costing horizon for error and control signal.
λ: is a control weighting sequence (often chosen as 0.5).

N_1 is often set to one, however, if the process time delay d is known, then N_1 is set to be equal to d. N_2, the maximum costing horizon is set to be equal to the process time response, or at least close to it.

In general, most models are stabilised with GPC values of 1 and 10 for N_1 and N_2, respectively. $\lambda(j)$ is often taken as a constant, λ and is time invariant. Now having, from equation (7.15), if N_1 is set to 1, the predicted process output sequence is giving by:

$$\begin{aligned} y(k+1) &= G_1\Delta u(k) + F_1 y(k) + E_1 e(k+1) \\ y(k+2) &= G_2\Delta u(k+1) + F_2 y(k) + E_2 e(k+2) \\ &\vdots \\ y(k+N_2) &= G_1\Delta u(k+N_2-1) + F_{N_2} y(k) + E_{N_2} e(k+N_2) \end{aligned}$$

The future predicted output $y(k+j)$ is constituted *in fine* of three terms:

- The first one depends on the future control actions (yet to be determined),
- The second term, depends on present output values depending on past controls,
- The second item depends also of filtered measured variables and one depending on future noise signals. Let $f(k+j)$ be that component of $y(k+j)$ composed of signals which are known at time k,

so that, for example,

$$\begin{aligned} f(k+1) &= [G_1(q^{-1}) - g_{10}]\Delta u(k) + F_1 y(k) \\ f(k+2) &= q[G_2(q^{-1}) - q^{-1}g_{21} - g_{20}]\Delta u(k) + F_2 y(k) \\ &\vdots \\ f(k+N_2) &= q^{N_2-1}[G_{N_2}(q^{-1}) - q^{-N_2+1}g_{N_2N} - \cdots - g_{N_20}]\Delta u(k) + F_{N_2} y(k) \end{aligned}$$

where $G_i(q^{-1}) = g_{i0} + g_{i1}q^{-1} + \cdots$

On the vector form, the above equations can be written as follows:

$$\hat{y} = G\tilde{u} + f \tag{7.23}$$

where the vectors $\in \mathbf{R}^{N_2}$.

$$\hat{y} = [\hat{y}(k+1), \hat{y}(k+2), \ldots, \hat{y}(k+N_2)]^T$$
$$\tilde{u} = [\Delta u(k), \Delta u(k+1), \ldots, \Delta u(k+N_2-1)]^T$$
$$f = [f(k+1), f(k+2), \ldots, f(k+N_2)]^T$$

The first j terms in $G_j(q^{-1})$ are the values of the system step-response parameters, and, therefore, $g_{ij} = g_j$ for $j = 0, 1, 2 \cdots < i$, is independent of the particular G polynomial. The matrix G is then lower-triangular of dimension $N_2 \times N_2$.

$$G = \begin{bmatrix} g_0 & 0 & \cdots & 0 \\ g_1 & g_0 & \cdots & 0 \\ . & .. & \cdots & . \\ \vdots & \vdots & & \vdots \\ g_{N_2-1} & g_{N_2-2} & \cdots & g_0 \end{bmatrix}$$

If the plant has a dead time equal to $d > 1$ and unknown, the first $d-1$ rows of G will be null. If d is known then, N_1 is set to equal d and the leading element of the matrix is non zero. However, even if d is unknown a stable solution is still achievable [53].

Using equation(7.23), the cost function expressed in equation(7.22), becomes:

$$J_1 = (y - w)^T (y - w) + \lambda \tilde{u}^T \tilde{u} \tag{7.24}$$

Minimising J_1, assuming no constraints on future controls, results in the projected control law:

$$\tilde{u} = (G^T G + \lambda I)^{-1} G^T (w - f) \tag{7.25}$$

The first element of $\tilde{u}$ is $\Delta u(k)$ so that the current control $u(k)$ is given by

$$u(k) = u(k-1) + \bar{g}^T (w - f) \tag{7.26}$$

where $\bar{g}^T$ is the first row of $(G^T G + \lambda I)^{-1} G^T$.

7.4 Nonlinear model predictive control

Even if most industrial processes are of nonlinear to a highly nonlinear nature, most implemented MPC applications still use linear MPC algorithms (i. e., based on linear dynamic models) [47, 55].

One of the reasons for the predominance of linear MPC is the widespread and good understanding of linear identification methods on top of having proven their efficiency in the past three decades for modelling and identification. However, the main justification of using linear models to describe nonlinear processes comes from the fact that many control problems aim to maintain the output process at a desired steady state (i. e., a regulation problem), rather than moving rapidly from one set-point to another (i. e., a pursuit problem). This is the case for a distillation column in refinery processing. If the milk plant described in Chapter 2 is constrained to production with pasteurised milk, then we are confronted with a regulation problem. In that case, a carefully identified linear model can be sufficiently accurate in the neighbourhood of a single operating point.

Nevertheless, in some cases nonlinear behaviours are significant and their effects on the process are substantial, which makes the use of a nonlinear prediction model justified despite extra modelling and control law determination efforts. Examples of such applications, where a linear model is unable to predict accurately the future process output(s), could be given by:

- regulator control problem where the process is highly nonlinear and is subject to large disturbances (e. g., pH control),
- servo control problem where the operating point changes frequently within a significant range.

In the literature, many ways to address nonlinear process behaviour using an approximation based on a linear dynamic model, have been described. A gain scheduling method, where model gains are obtained at each control iteration by applied perturbations, a nonlinear steady state model was used with a DMC algorithm given in [56]. Garcia (1984) [54], proposed an extension of QDMC to nonlinear processes called NQDMC (N for nonlinear). Gattu and Zafiriou (1992) [64] applied the concept of QDMC and modified it using Kalman's filter, in order to handle open loop unstable systems. A nonlinear adaptation of GPC has also been investigated in [58]. However, optimal control can only be achieved using a true nonlinear model of the plant instead of having a linearised approximation. MPC algorithms using a nonlinear model are referred to as nonlinear MPC (NMPC).

As opposed to linear MPC where DMC, GPC, IDCOM are considered as standard algorithms, Nonlinear MPC (NMPC), does not (at least not yet) posses its own standard algorithms and formulations. The main cause is once again the relative youth of NMPC and the fact that a wide range of model types as well as optimisers can be used. However, several companies established their own NMPC algorithms. Qin and Badgwell [57], enlighten the readers and provided a description of some of the most successful NMPC algorithms developed by several companies, e. g., Aspen Target from Aspen Technology using a nonlinear state space model, MPC from Continental Controls using a static nonlinear polynomial model and Process Perfector from Pavilion Technology using ANN-based models.

As machine learning took over wide ranges of data based modelling over physical modelling, the use of ANN in model predictive control increased in popularity, and an overview of the few existing applications of MPC using ANN as internal models is given in Section 7.5.

The basis of a neural predictive controller NMPC is shown in the diagram Figure 7.3. The MPC is here based on an ANN model, used to obtain predictions to be used with for the calculation of the errors between the real and modelled process outputs within the optimization algorithm. The control variable u is obtained by minimising a criterion function J, where $N1$ is the time delay given in number of samples (if any), N the prediction horizon and $N2$ the control horizon:

$$J_{=} \sum_{i=N1}^{N1+N} [y_r(k+i) - \hat{y}(k+i|k)]^2 + \sum_{i=0}^{i=N2} \lambda[u(k+i)]^2 \tag{7.27}$$

At each instant k, the predicted output $\hat{y}(k+i|k)$ is compared to a reference $y_r(k+i)$ describing the optimal trajectory taken to reach the target, in the regulation case a constant C. J is the criterion to be minimised and is given by equation (7.27).

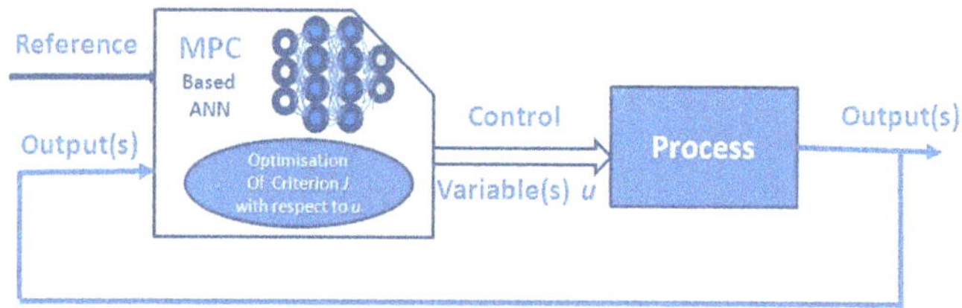

Figure 7.3: Neural MPC configuration.

As an analytical solution is not possible, a numerical one may be obtained using an optimiser in order to obtain the value of u that minimises J, equation (7.27). The practitioner are provided with a large number of optimisation routine to choose among in the literature. As an example, the interior-point method based optimiser for use with MPC is proven more efficient [78].

7.5 Case study: neural networks and model predictive control for pasteurisation temperature control in a plate heat exchanger

This section investigates the use of Artificial Neural Networks (ANNs), more precisely multi-layer perceptrons (MLPs), for the nonlinear modelling and predictive control of a milk pasteurisation plant temperature based on Plate Heat Exchanger (PHE).

As explained earlier, MPC algorithms require the development of a predictive model to predict future process outputs. For the sake of the internal model and using data gathered from an industrial milk plant, a nonlinear multi-step ahead ANN predictive model is designed and developed. A Neural Model Predictive Controller (NMPC) will then be designed on the same basis for the control of milk pasteurization temperature. Simulation results are presented and conclusions are drawn.

Previous works have been done on modelling and control of milk pasteurization using predictive control by the same author and others. In [68], a first principle physical model has been developed as an internal model of a Predictive Functional Controller (PFC) where it is shown to perform better than a classical PID; see [68, 69] assuming that the linear model obtained from first principles modelling is valid.

The fact that the milk pasteurization plant operates around a fixed set-point made the linearity assumption valid. For a wider control range, a model well able to predict future outputs over a wider temperature range is needed and a linear model cannot achieve such a performance.

Well-established and available analytical MPC strategies are restricted to linear models in order to compute an analytical control law [53, 49]. Therefore, for industrial processes, a linearised model is most often used as an internal model.

Indeed, a full nonlinear complete model of any industrial processes can be obtained via physical modelling. However, this is not an easy task due to most processes physical complexity and high nonlinearity, and makes the task not trivial and very time and efforts consuming.

The usage of ANNs finds here all its justification, due to their ability (see Section 2.5) to search for a valid nonlinear model obtained from input/output data. The validated nonlinear model can then be used, as an internal model, for the implementation of a “nonlinear” neural predictive controller (NMPC). The extra computation involved in NMPC is largely achievable within the 12-second sampling period of the pasteuriser data collection.

7.5.1 Milk pasteurisation process

The studied plant is a pasteurisation plant based on a Plate Heat Exchanger (PHE) from the Alpha Laval Clip range the Clip 10-RM. PHEs are efficient heat exchangers consisting in packs of corrugated plates clamped in a frame. The corrugations and plate design favor incising flow turbulence of the liquids (in this case milk and water for product and medium, respectively) [70].

The Clip 10-RM is constituted of five sections namely, S1 to S5, of different sizes (number of plates), dedicated to: Heating (Sections S1 and S3), regeneration (Sections S4 and S2) and cooling (Section S5). The treatment operations details are depicted in Figure 7.4(a), and summarised in what follows.

The raw milk in introduced to the PHE with a usual concentration of around 4.1 % in fat content, in Section S4. The raw milk must be transported at a regulated temperature of 2.0 °C, prior to be fed to the PHE. In Section 4, the raw milk in preheated by regeneration (using the already pasteurised milk coming from heating sections, to

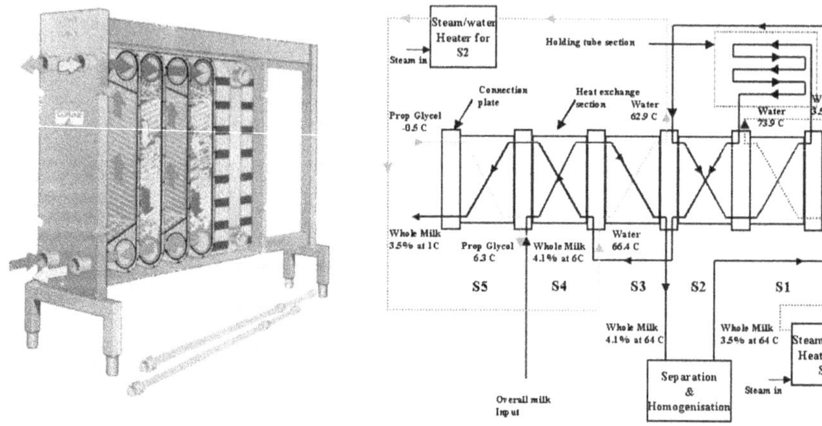

(a) Schematic of the pasteuriser (PHE)

(b) Block diagram and representation

Figure 7.4: Plate heat exchanger schematics representation.

reach a temperature of 60.5 °C, where, as a result, the outgoing pasteurised milk temperature is reduced to 11.5 °C.

Entering Section S3, the milk now at 60.5 °C, is heated to a regulated temperature of 64.5 °C using hot water as a heating medium via a control loop. At this stage, the milk is separated from the fat, then standardised and homogenised to a concentration of 3.5 %, before being fed to Section S2. The homogenised milk is then heated by regeneration to a temperature of 72 °C using the already pasteurised milk as a medium. Only then, the milk can be brought to the actual regulated pasteurisation temperature of 75.0 °C in Section S1, using hot water already heated at a temperature of 77.0 °C as a medium. In order to respect the pasteurisation specifications, the heated milk at the output of Section S1 is held at the "pasteurisation" temperature of 75.0 °C for 15 s in the holding tube section. Note that pasteurisation temperature is 72 °C [65], but is set here at 75.0 °C, in order to avoid a temperature drop out of pasteurisation range in case of disturbances or control glitches. Once the milk is "pasteurised", the cooling process may begin and start in Section S4, then S2 to finally be brought from a temperature of 11.5 °C at the output of section S4 to a temperature of around 1.0 °C using propylene glycol as a medium at a temperature of −0.5 °C in Section S5.

Note that in the heating regulated sections S3 and S1, the milk is heated using hot a water medium, that is itself heated in steam/water heater of type CB76 from Alfa Laval. The water temperature may be considered, as well as the steam flow injected into boilers to heat water, as the manipulated or control variable for milk pasteurisation. Figure 7.4(b), shows how the milk pasteurisation temperature may be originally, a function of three inputs: steam flow injected in steam/water heater 1, steam flow injected in steam/water heater 2 and the milk input temperature, labelled as F_{v1}, F_{v2}, and T_{im}, respectively. The milk pasteurisation temperature may be represented by a Multi-Input Single Output (MISO) system, with F_{v1}, F_{v2} and T_{im} as inputs and y, the milk pasteurisation temperature, as output.

7.5.2 Justification and rationale for using MPC

The set-point temperature of 75.0 °C ensures pasteurization and prevents the milk temperature from dropping under 72.0 °C. This implies a security margin of 3.0 °C to prevent any negative temperature variance around the set-point exceeding 3.0 °C. However, there are two major disadvantages to this approach:

- wastage of energy occurs when heating the milk at 75.0 °C, where the pasteurisation temperature is only 72.0 °C, and
- positive temperature variance is as likely to occur than negative temperature variance. This could bring the milk up to 78.0 °C, in which case the milk nutrient value may be altered as proteins are more likely to burn. This could also affect the taste of the milk, giving it a burnt taste.

The main goal of an advanced control strategy for controlling the milk pasteurization temperature is to reduce the variance in milk temperature. Although, controlling the temperature variance will ensure that the nutrient value of the milk is not affected, it will not prevent energy loss. In order to save energy, the pasteurisation target temperature has to be reduced. This may be possible after the variance is reduced. For example, if the variance is brought to 1.0 °C, then the target pasteurisation temperature may be set to 73.5 °C, still keeping a security margin of 0.5 °C. The "squeeze the variance shift the target" approach can be shown in Figure 7.5. Although the reduction of 1.0–2.0 °C in the set-point temperature may seem a small change; it is important in the context of liquid milk not being a large profit product. The smallest reduction in the set-point temperature is, therefore, a relatively important improvement when considering the entire production time. MPC has been successful as an advanced control approach for

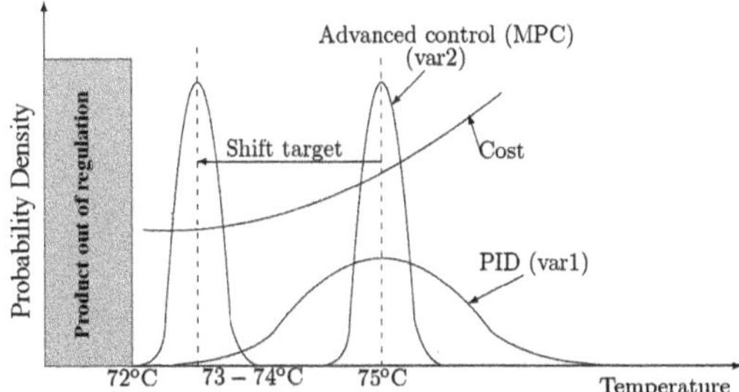

Figure 7.5: Goal of an advance control approach: Squeeze the variance shift the target.

more than 2000 industrial applications [47], and has been chosen in order to improve the control of milk pasteurisation.

7.5.3 Modelling of the pasteurization plant using ANN

As seen in Section 3.5, Cybenko [71] stated that one layer back-propagation neural networks can approximate any nonlinear function, generating complex decision regions for input-output mapping. In this case, a multi-layer network with singles input and output layers and two hidden layers is used as a modelling tool. The advantages of using more than one hidden layer are stated in Section 3.5.

In order to perform meaningful training of the ANN, the inputs are chosen to be a sequence of data gathered from, F_{v1}, F_{v2}, and eight delayed values of the output signal y; see Figure 7.4(b). The choice of the input regressor parameters is heavily influenced by the a priori information gathered from the first principles physical model used in [68]. The milk temperature in [68], is modelled by an eighth-order linear system, justifying the use of eight delayed signals of $\hat{y}(k)$ in equation (7.28). Note that the input milk temperature T_{im} is not used as part of the input data of information. The reason for that is that its relatively constant temperature of 2 °C does not hold any informative value and at most can be considered as an input disturbance.

Selecting the optimal or most adequate ANN topology is nor an easy nor a straightforward task, and ANNs remains a heuristic approach, and no rules or theorems exist for the matter. However, many approaches both intuitive and computational, may be used to approach an optimal topology or at least find an adequate one. As stated in Section 3.9.4, network growing and network pruning may be applied. Optimal Brain Damage (OBD) developed by Le Cun [72] is a pruning computational approach that starts with a sufficiently big topology; the ANN is pruned by eliminating the links containing smaller weights using a *weight elimination method.* Note for this application case of pasteurisation control, no computational topology optimisation is needed and the network topology is intuitively chosen based on first principle modelling and is shown in Figure 7.6. The ANN neurons in the two hidden layers are chosen to be tangent sigmoidal neurons, where the output layer neuron is a linear neuron. The prediction will be given by the ANN obtained after appropriate training on the form:

$$\hat{y}(k) = \text{ANN}\left[\hat{y}(k-1), \hat{y}(k-2) \cdots \hat{y}(k-8), F_{v1}(k-1), F_{v1}(k-1), F_{v2}(k-1), F_{v2}(k-1)\right] \quad (7.28)$$

where the ANN function variables are the matrix of weight and biases.

Alternatively, starting with a small enough ANN and increasing the hidden layers neurons numbers until reaching a size well able to translate the process complexity, giving a good prediction model. The modelling approach chosen in this section is in its large lines similar to the one described by Nõrgaard et al. in [73], and the network was trained for 2000 epochs using a set of data that consists of data subsets obtained

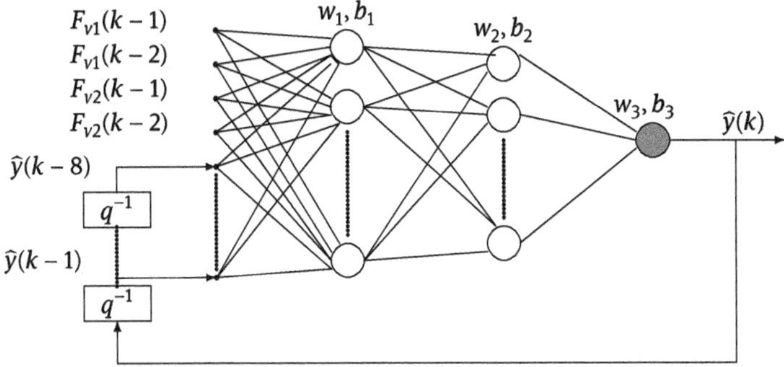

Figure 7.6: ANN topology and input signals used for training.

during a series of test protocols, where F_{v1} and F_{v2} were varied around the operating region. In a cross-validation approach, four subsets of data were used i turn. Three subsets and a separate subset was used for raining and validation, respectively, in order to maximise validity chances. The method prove to be useful when the size of the sample space is constrained and limited. Moreover, having several validation estimates covering the entire training data set gives a better confidence degree to the estimates.

To avoid overtraining, i. e., deterioration of the model as it tries to fit the training set, a monitored training is performed checking at each epoch the Sum Squared Error (SSE) of the validation set along with the SSE of the training set. Overtraining and its effect are explained in detail in [74]. Several training tries and runs are performed and the best results for two validation subsets are presented in Figure 7.7.

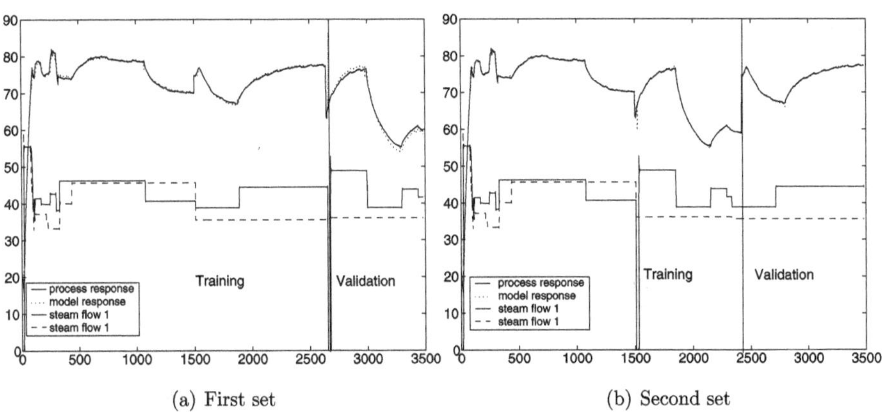

Figure 7.7: Training and validation of the ANN model.

In the figure, the process and ANN model's response to a change in F_{v1} and F_{v2} flows given in m^3/s (shown scaled on the graph by a factor of 500) can be seen for the training and validation data sets.

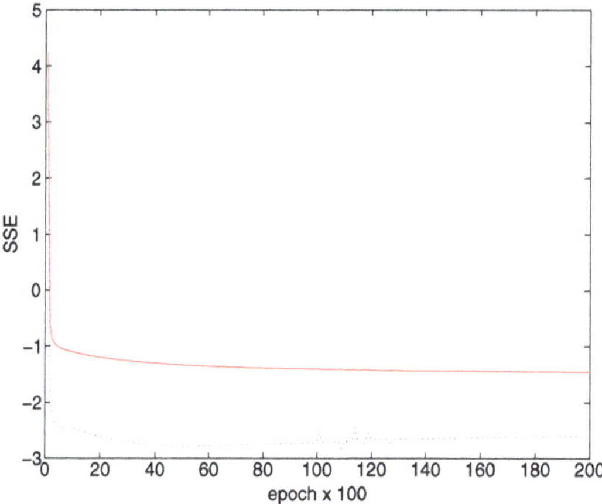

Figure 7.8: Validation SSE progression versus training epoch number.

On the other hand, and for the first set, the validation SSE expressed at every epoch is given in Figure 7.8. It is clearly seen that after epoch 55, the SSE start increasing depicting the occurrence of overtraining, the model parameters are saved then to keep the equivalent model. For the obtention of accurate and meaningful results, reaching then a global or at least decent local minimum, a number of simulation runs are performed for every validation subset. The training operation terminated four ANN models namely ANN_{M1}, ANN_{M2}, ANN_{M3}, and ANN_{M4} are obtained using the four validation subsets, in turn, while the rest of the data is used for training and the minimum validation SSEs obtained and are displayed in Table 7.7 along with the corresponding epoch number.

Table 7.7: ANN Training and validation errors.

MAE/Data subsets	ANN_{M1}	ANN_{M2}	ANN_{M3}	ANN_{M4}
MAE_t	0.2378 °C	0.3230 °C	0.1351 °C	0.2558 °C
MAE_v	0.5418 °C	0.2040 °C	1.0912 °C	0.4689 °C
Epoch stop	5000	5500	2800	11000

Conception ad training of the ANN models was performed using the Matlab artificial neural network toolbox [75], and more precisely the celebrated propagation training algorithm described in detail in [76]. The final overall multi-model is given by a linear weighted combination of the four obtained models ANN_{M1} to ANN_{M4} as shown in Figure 7.9.

The weights of the linear combiner are determined using a simple least square method, training the overall linear combiner using all four data sets. The weights W_{M1},

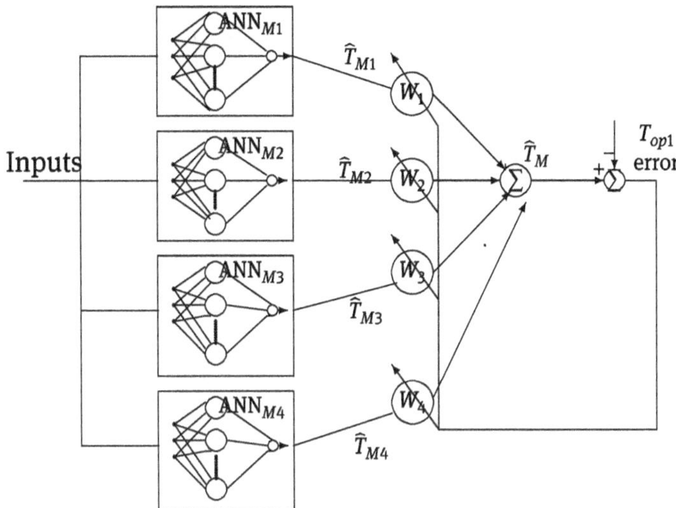

Figure 7.9: Overall ANN model: linear combiner.

W_{M2}, W_{M3}, and W_{M4} are found using a simple least-squares method given in equation (7.29).

$$\begin{bmatrix} W_{M1} \\ W_{M2} \\ W_{M3} \\ W_{M4} \end{bmatrix} = \left(\sigma^T \sigma\right)^{-1} \sigma^T \begin{bmatrix} T_{\text{om1}}(1) \\ T_{\text{om1}}(2) \\ \vdots \\ T_{\text{om1}}(n) \end{bmatrix} \tag{7.29}$$

where

$$\sigma = \begin{bmatrix} \hat{T}_{M1}(1) & \hat{T}_{M2}(1) & \hat{T}_{M3}(1) & \hat{T}_{M4}(1) \\ \hat{T}_{M1}(2) & \hat{T}_{M2}(2) & \hat{T}_{M3}(2) & \hat{T}_{M4}(2) \\ \vdots & \vdots & \vdots & \\ \hat{T}_{M1}(n) & \hat{T}_{M2}(n) & \hat{T}_{M3}(n) & \hat{T}_{M4}(n) \end{bmatrix}$$

With T_{om1}, is the temperature of the output medium (i. e., pasteurised milk) at the output of Section S1 prior to holding, and $\hat{T}_{M2}(i)$ is the output of each individual ANN model.

Once the weights of each linear combiner are determined, using equation (7.29), the overall model ANN$_P$ is given in Figure 7.10.

Combining all four different models gives the overall ANN model, ANN$_P$, better generalization capability [66].

It can be seen that the model response $\hat{T}_P$ follows closely the process output T_{om1} over the entire data sets 1, 2, 3, and 4. Moreover, it can be seen from Figure 7.10 that the model responses are enclosed within the sensor's uncertainty bounds regardless of

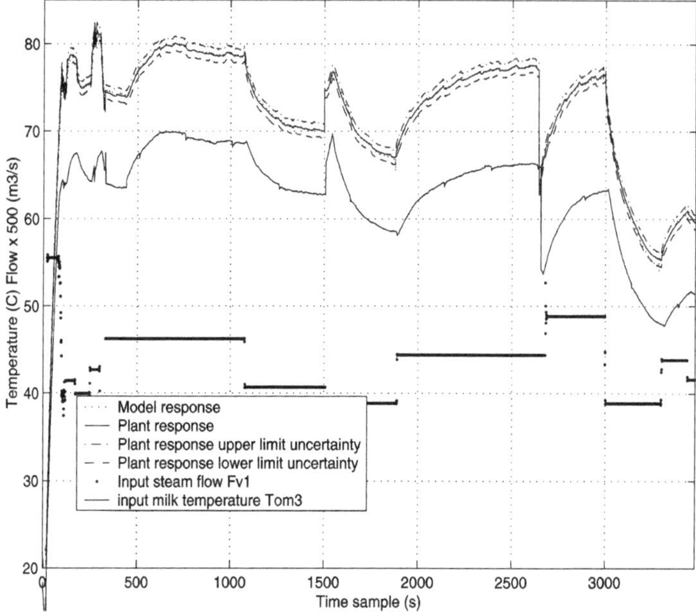

Figure 7.10: Final ANN_P response, resulting from linear combination.

the region of operation, which makes further improvement of the model unnecessary unless more accurate sensors are fitted.

The obtained results, in terms of MAE for the validation subsets, along with the optimal linear weights, are summarised in Table 7.8:

- MAE_V values for the three linear combiners,
- mean MAE_V's obtained from cross model selection,
- the benchmark MAE given by the linear model obtained in [67],
- W_i are the linear combiner's weights.

Table 7.8: Linear combiners results.

ANN modelling results						
Model	**Mean MAE_V °C**	**Combiner MAE_V °C**	**Combiner's weights**			
			W_1	W_2	W_3	W_4
ANN_P	0.2535	0.2187	0.3127	0.3054	0.3823	0.0001
Linear first principles modelling results [67]						
Model			MAE_V °C			
Past. process			0.6349			

7.5.4 Neural predictive controller design

Now that a validated nonlinear ANN model of the pasteurization process is established, a neural predictive controller NMPC is designed accordingly as shown in Figure 7.11. The ANN model obtained in Section 7.5.3 is used as an internal model in order to produce prediction data for the optimization algorithm. The control variable u is obtained by minimising a criterion function J, where, $N1$ is the time delay (if any), N and $N2$ being the prediction and control horizon, respectively.

$$J = \sum_{i=N1}^{N1+N} [y_r(k+i) - \hat{y}(k+i|k)]^2 + \sum_{i=0}^{i=N2} \lambda[u(k+i)]^2 \tag{7.30}$$

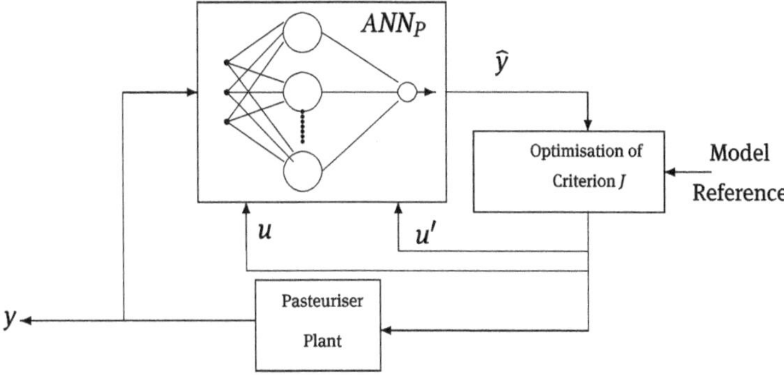

Figure 7.11: Structure of the neural NMPC controller.

At each instant k, the predicted output temperature $\hat{y}(k+i|k)$ is compared to a reference trajectory $y_r(k+i)$ representing the desired trajectory to reach the target. The control variable u represents physically by the steam flow F_{V1}, and steam flow F_{V3} used to heat the medium that is in turn used to heat the milk in Section S3, represented in Figure 7.12 and considered as an input disturbance in the control of pasteurisation temperature in Section S1.

An optimiser has to be used in order to obtain the value of u that minimises J. The optimiser used in this case study is based on a gradient decent method and is available in the NAG toolbox used with Matlab [77]. By applying different inputs to the ANN model, shown in Figure 7.11 represented by u' until finding the best value of u that minimises J.

More efficient optimisers can be used and a large number of optimisation routines are available in the literature. As an example, the interior-point method based optimiser for use with MPC is proven more efficient [78]. However the optimiser used for this case study, is found to give satisfactory results due to the relatively slow response time of the process (i. e., 12 s), making the convergence speed of the optimisation not an issue to consider.

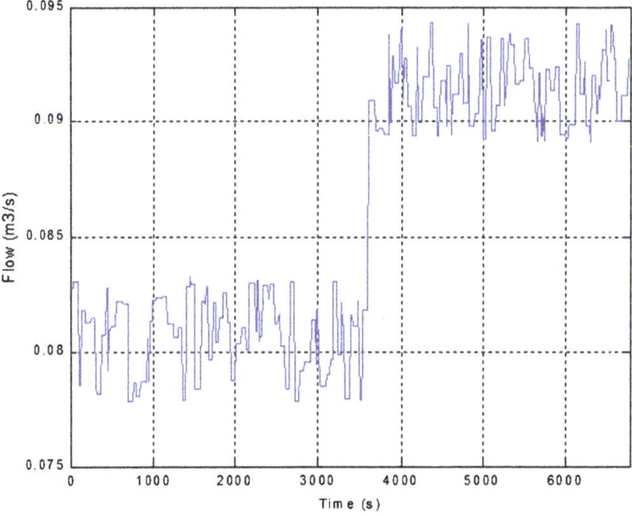

Figure 7.12: Input disturbance.

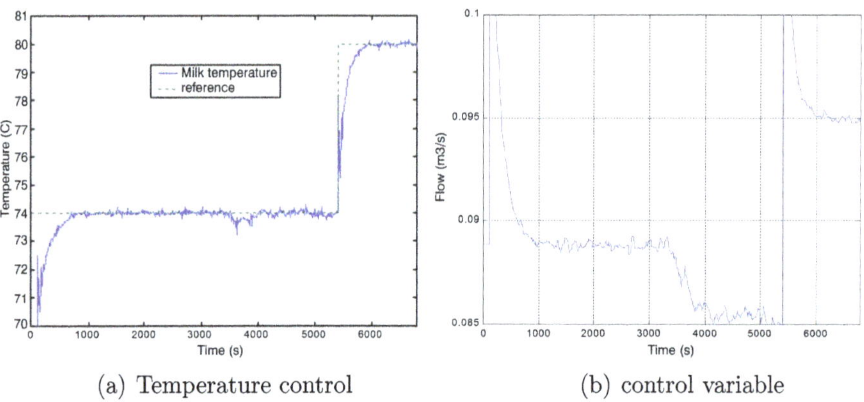

Figure 7.13: NMPC behaviour to disturbance and set-point changes.

A prediction horizon $N = 25$, is chosen while a single step control horizon of $N2 = 1$ is applied as in practice such control action is sufficient. Figure 7.13(a) shows the results obtained using NMPC for a reference pasteurisation temperature of 74 °C for the first control sequence.

Then the temperature target for testing the controller and model compliance with other temperature set-points is brought up to 85 °C.

Indeed, even if milk pasteurisation is performed usually at around 74/75 °C, other heating processes, such as high temperature short time pasteurisation, thermisation, etc. may use a different set-point or regulation temperature.

Moreover, during the control process he process is subject to input disturbances given by F_{v3} and $T_{\rm im}$ the milk input temperature, therefore, the model predictions given by the NMPC internal model needs to be tested accordingly during simulations. The input disturbances in terms of variance and sudden changes are shown in Figure 7.12. As can be seen, the value of F_{v3} varies in a pseudo random manner around a given value, in reality around 0.075 m^3/s, in order to regulate the intermediate temperature $T_{\rm om3}$, the milk temperature at the output of Section S3 that must be kept at the separation temperature of 64 °C. The large change at time at 3, 500 s in the value of F_{v3} is improbable to happen in real life, yet has been created to test the capabilities of the NMPC in extreme conditions of a sudden drop/increase of steam flow.

As a general conclusion, the nonlinear MPC, based on an ANN internal model is found to perform better than linear MPC approaches namely: PFC and GPC especially over a wide temperature range. The rationale of the superiority of NMPC is due to the nonlinear nature of the internal ANN model able to better matches the process output away from a given set-point around which a linear internal model would have been tuned. The main drawback of NMPC, is that it requires a numerical solution due to the inherent nonlinearity of the internal model, and is obtained by running an optimisation routine, needing therefore more computational efforts. The issue become sensitive when dealing with processes with fast responses nd the sampling period must be long enough to permit the execution for an optimisation run, as this is repeated at each sample time.

In the studied case, the process is relatively slow and 1 s sampling period is comfortable enough to implement most optimisation routines. Even though NMPC seems to give the better results, implementing such a relatively complicated technique is only justifiable if the process is subject to frequent set-point changes over a wide temperature range. In the case of pasteurisation, where the set-point is fixed, such a control approaches is not justified. In order to asses the benefits of NMPC over classical and linear MPC for milk pasteurisation temperature control, the reader is directed to the works of the same author [3, 4] and more recently a comparison of advanced control strategies for the control of milk pasteurisation in plate heat exchangers [79]. Numerical assessment of benchmark MPC and PID controllers along with the proposed NMPC approach are presented and the superiority of NMPC control, especially in the case of set up change, is clearly highlighted.

Bibliography

[1] Roupas, P. (2008) Predictive modelling of dairy manufacturing processes. International Dairy Journal, 18(7), 741–753. doi:10.1016/j.idairyj.2008.03.009.

[2] Paquet, J., Lacroix, C., and Thibault, J. (2000) Modeling of pH and acidity for industrial cheese production. Journal of Dairy Science, 83, 2393–2409.

[3] Khadir, M. T. and Ringwood, J. V. (2003) Application of generalised predictive control to a milk pasteurisation process, in Hamza, M. H. (ed.), Salzburg. Proceedings of the IASTED International Conference on Intelligent Systems and Control, pp. 230–235.
[4] Khadir, M. T. and Ringwood, J. V. (2003) Linear and nonlinear model predictive control design for a milk pasteurization plant. Control and Intelligent Systems, 31(1), 37–44.
[5] Funes, E., Allouche, Y., Beltrán, G., and Jiménez, A. (2015) A review: artificial neural networks as tool for control food industry process. Journal of Sensor Technology, 5, 28–43.
[6] Arumugasamy, S. K. and Ahmad, Z. (2009) Elevating model predictive control using feedforward artificial neural networks: a review. Chemical Product and Process Modeling, 4(1).
[7] Lawrynczuk, M. (2008) Modelling and nonlinear predictive control of a yeast fermentation biochemical reactor using neural networks. Chemical Engineering Journal, 145(2), 290–307. doi:10.1016/j.cej.2008.08.005.
[8] Zhan, J. and Ishida, M. (1997) The multi-step predictive control of nonlinear SISO processes with a neural model predictive control (NMPC) method. Computers & Chemical Engineering, 21(2), 201–210.
[9] Bhat, N. and McAvoy, T. (1990) Use of neural nets for dynamic modeling and control, of chemical processes. Computers & Chemical Engineering, 14, 573–583.
[10] Lee, M. and Park, S. (1992) A new scheme combining neural feedforward control with model-predictive control. AIChE Journal, 3(2), 193–200.
[11] Willis, M. J., Montague, G. A., Di Massimo, C., Morris, A. J., and Tham, M. T. (1991) Artificial neural networks and their application in process engineering, in IEEE Coll. on Neural Networks for Systems: Principles and Applications, London.
[12] Declercq, F. and De Keyser, R.. (1996) Comparative study of neural predictors in model based predictive control, in Proc. of International Workshop on Neural Network for Identification, Control, Robotics, and Signal/Image Processing, Bratislava, pp. 20–28.
[13] Hernandez, E. and Arkun, Y. (1990) Neural network modeling and an extended DMC algorithm to control nonlinear systems, in Proc. American Control Conf., San Diego, pp. 2454–2459.
[14] Kondakci, T. and Zhou, W. (2017) Recent applications of advanced control techniques in food industry, Food and Bioprocess Technology 10, 522–542. doi:10.1007/s11947-016-1831-x.
[15] Paquet-Durand, O., Solle, D., Schirmer, M., Becker, T., and Hitzmann, B. (2012) Monitoring baking processes of bread rolls by digital image analysis. Journal of Food Engineering, 111(2), 425–431.
[16] Linko, P., Uemura, K., Zhu, Y. H., and Eerikainen, T. (1992) Application of neural network models in fuzzy extrusion control. Trans IChemE, 70(C3), 131–137.
[17] Yüzgeç, U., Becerikli, Y., and Turker, M. (2008) Dynamic neural-networkbased model-predictive control of an industrial baker's yeast drying process. IEEE Transactions on Neural Networks, 19(7), 1231–1242.
[18] Köni, M., Yüzgeç, U., Türker, M., and Dinçer, H. (2010) Adaptive neurofuzzy-based control of drying of baker's yeast in batch fluidized bed. Drying Technology, 28(2), 205–213.
[19] Cubillos, F. A., Vyhmeister, E., Acuña, G., and Alvarez, P. I. (2011) Rotary dryer control using a grey-box neural model scheme. Drying Technology, 29(15), 1820–1827.
[20] Torrecilla, J. S., Otero, L. and Sanz, P. D. (2005) Artificial neural networks: a promising tool to design and optimize high-pressure food processes. Journal of Food Engineering, 69, 299–306. doi:http://dx.doi.org/10.1016/j.jfoodeng.2004.08.020.
[21] Mete, T., Ozkan, G., Hapoglu, H., and Alpbaz, M. (2012) Control of dissolved oxygen concentration using neural network in a batch bioreactor. Computer Applications in Engineering Education, 20(4), 619–628. doi:https://doi.org/10.1002/cae.20430.
[22] Eerikäinen, T., Linko, P., Linko, S., Siimes, T., and Zhu, Y. H. (1993) Fuzzy logic and neural network applications in food science and technology. Trends in Food Science & Technology, 4(8), 237–242.

[23] DiMassimo, C., Lant, P. A., Saunders, A., Montague, G. A., Tham, M. T., and Morris, A. J. (1992) Bioprocess applications of model-based estimation techniques. Journal of Chemical Technology and Biotechnology, 53(3), 265–277.

[24] Royce, P. N. (1993) Discussion of recent developments in fermentation monitoring and control from a practical perspective. Critical Reviews in Biotechnology, 13(2), 117–149.

[25] Chtourou, M., Najim, K., Roux, G., and Badhhou, B. (1993) Control of a bioreactor using a neural network. Bioprocess Engineering, 8(5–6), 251–254.

[26] Zhu, Y. H., Rajalahti, T., and Linko, S. (1996) Application of neural networks to lysine production. Chemical Engineering Journal and the Biochemical Engineering Journal, 62(3), 207–214.

[27] Linko, S. and Linko, P. (1998) Developments in monitoring and control of food processes. Food and Bioproducts Processing: Transactions of the Institution of Chemical Engineers, Part C, 76(3), 127–137.

[28] Patnaik, P. R. (1997) A recurrent neural network for a fed-batch fermentation with recombinant Escherichia coli subject to inflow disturbances. Process Biochemistry, 32(5), 391–400.

[29] Wang, Y., Luopa, J., Rajalahti, T., and Linko, S. (1995) Strategies for the production of lipase by Candida rugosa; neural estimation of biomass and lipase activity. Biotechnology Techniques, 9(10), 741–746.

[30] Popescu, O., Popescu, D., Wilder, J., and Karwe, M. (2001) A new approach to modelling and control of a food extrusion process using artificial neural network and an expert system. Journal of Food Process Engineering, 24, 17–36. doi:http://dx.doi.org/10.1111/j.1745-4530.2001.tb00529.x.

[31] Simutis, R., Havlik, I., and Lubbert, A. (1992) Fuzzy-aided neural network for real-time state estimation and process prediction in the alcohol formation step of production-scale beer brewing. Journal of Biotechnology, 27(2), 203–215.

[32] Latrille, E., Corrieu, G., and Thibault, J. (1994) Neural network models for final process time determination in fermented milk production. Computers & Chemical Engineering, 18(11–12), 1171–1181.

[33] Kosola, A. and Linko, P. (1994) Neural control of fed-batch baker's yeast fermentation. Developments in Food Science, 36, 321–328.

[34] Petre, E., Sendrescu, D., and Selisteanu, D. (2011) Neural networks based model predictive control for a lactic acid production bioprocess, in König, A., Dengel, A., Hinkelmann, K., Kise, K., Howlett, R. and Jain, (eds.) Knowledge-Based and Intelligent Information and Engineering Systems, Lecture Notes in Computer Science, vol. 6884. Springer Berlin Heidelberg, pp. 388–398.

[35] O'Farrell, M., Lewis, E., Flanagan, C., Lyons, W. B., and Jackman, N. (2004a) Employing spectroscopic and pattern recognition techniques to examine food quality both internally and externally as it cooks in an industrial oven, in: Lopez-Higuera, J. M. and Culshaw, B. (eds.), Santander. Second European Workshop on Optical Fibre Sensors, EWOFS'04. pp. 313–316.

[36] Benne, M., Grondin-Perez, B., Chabriat, J. P., and Hervè, P. (2000) Artificial neural networks for modelling and predictive control of an industrial evaporation process. Journal of Food Engineering, 46(4), 227–234.

[37] Pitteea, A. V., King, R. T. F., and Rughooputh, H. C. S. (2004) Intelligent controller for multiple-effect evaporator in the sugar industry, in Industrial Technology, 2004. IEEE ICIT '04. 2004 I.E. International Conference on, 8--10 Dec. 2004, pp. 113, 117–182.

[38] Vasičkaninovã, A., Bakošová, M., Mèszàros, A., and Klemeš, J. J. (2011) Neural network predictive control of a heat exchanger. Applied Thermal Engineering, 31(13), 2094–2100.

[39] Martynenko, A. I. and Yang, S. X., (2007) An intelligent control system for thermal processing of biomaterials, in Networking, Sensing and Control. IEEE International Conference on, 15–17 April 2007, pp. 93–98.

[40] Mjalli, F. S. and Al-Asheh, S. (2005) Neural-networks-based feedback linearization versus model predictive control of continuous alcoholic fermentation process. Chemical Engineering & Technology, 28(10), 1191–1200.

[41] Escaño, J., Bordons, C., Vilas, C., García, M. R., and Alonso, A. A. (2009) Neurofuzzy model based predictive control for thermal batch processes. Journal of Process Control, 19(9), 1566–1575.

[42] O'Farrell, M., Lewis, E., Flanagan, C., Lyons, W. B., and Jackman, N. (2004) Using a reflection-based optical fibre system and neural networks to evaluate the quality of food in a large-scale industrial oven. Sensors and Actuators. A, Physical, 115(2–3 SPEC. ISS), 424–433.

[43] Riverol, C., Ricart, G., Carosi, C., and Di Santis, C. (2008) Application of advanced soft control strategies into the dairy industry. Innovative Food Science & Emerging Technologies, 9(3), 298–305.

[44] Kupongsak, S. and Tan, J. (2006) Application of fuzzy set and neural network techniques in determining food process control set points. Fuzzy Sets and Systems, 157(9), 1169–1178.

[45] Mesin, L., Alberto, D., Pasero, E., and Cabilli, A. (2012) Control of coffee grinding with artificial neural networks, in The 2012 International Joint Conference on Neural Networks (IJCNN), pp. 1–5.

[46] Cutler, C. R. and Ramaker, P. S. (1980) Dynamic matrix control – a computer algorithm, in Proceeding of the Joint Automatic Control Conference.

[47] Qin, S. J. and Badgwell, T. A. (2003) A survey of industrial model predictive control technology, Control Engineering Practice 11(7), 733–764.

[48] Piaget, J. and Inhelder, B. (1958) Growth of Logical Thinking. London: Routledge and Kegan Paul.

[49] Richalet, J. (1993) Pratique de la Commande Predictive. Paris: Traite des Nouvelles Technologies, Serie Automatique, Hermes.

[50] Mohtadi, C., Shah, S. L., and Clarke, D. W. (1992) Generalised Predictive Control for Multivariable Systems. Report No OUEL 1640/89, Oxford University, Department of Engineering Science.

[51] Zhang, R., Xue, A., Wang, S., Zhang, J., and Gao, F. (2012) Partially decoupled approach of extended non-minimal state space predictive functional control for MIMO processes. Journal of Process Control, 22(5), 837–851.

[52] Oblak, S. and Skrjanc, I. (2005) Multivariable fuzzy predictive functional control of a MIMO nonlinear system, in Proceedings of the 2005 IEEE International Symposium on, Mediterrean Conference on Control and Automation Intelligent Control, pp. 1029–1034.

[53] Clarke, D. W., Mohtadi, C., and Tuffs, P. S. (1987), Generalised predictive control-part I. The basic algorithm. Automatica, 23(2), 137–148.

[54] Garcia, C. E. (1984) Quadratic dynamic matrix control of nonlinear processes: an application to a batch reaction process, in AIChe Meeting, Houston Texas.

[55] Morari, M. and Lee, J. H. (2000) Model predictive control: past present and future. Computers & Chemical Engineering, 23, 667–682.

[56] Prett, D. M. and Gillette, R. D. (1980) Optimisation and constrained multivariable control of a catalytic cracking unit, in Proceeding of the Joint Control Conference.

[57] Qin, S. J. and Badgwell, T. A. (2000) An overview of nonlinear model predictive control applications, in Allgower, F., and Zheng, A. (eds.) Nonlinear Model Predictive Control. Switzerland: Birkhauser.

[58] Chen, W. H., Ballance, D. J., and Gawthrop, P. J. (1999) Nonlinear generalised predictive control and optimal dynamical inversion control, in IFAC World Congress, Beijing.

[59] Ricker, N. L., Subrahmanian, T., and Sim, T. (1988) Case studies of model predictive control in pulp an paper production, in Proceeding of the IFAC Workshop on Model Based Process Control, Oxford: Pergamon Press, pp. 13–22.

[60] Muske, K. R. and Rawlings, J. B. (1993) Model predictive control with linear models, AIChE Journal, 39(2), 262–287.
[61] Rawlings, J. B., Meadows, E. S., and Muske, K. R. (1994) Nonlinear model predictive control: a tutorial and survey, in ADCHEM Proceedings, Kyoto, Japan.
[62] Morari, M. and Lee, J. H. (2000) Model predictive control: past present and future. Computers & Chemical Engineering, 23, 667–682.
[63] Maciejowski, J. M. (2002) Predictive Control With Constraints. England: Pearson Education Limited.
[64] Gattu, G. and Zafiriou, E. (1992) Nonlinear quadratic dynamic matrix control with state estimation. Industrial & Engineering Chemistry Research, 31, 1096–1104.
[65] Holsinger, V. H., Rajkowski, K. T., and Stabel, J. R. (1997) Milk pasteurisation and safety: a brief history and update. Revue Scientifique Et Technique - Office International Des épizooties, 16(2), 441–466.
[66] Cloarec, G. M. (1998) Statistical methods for neural networks prediction models. Research report EE/JVR/97/2, Control System Group, School of Electronic Engineering, Dublin City University.
[67] Khadir, M. T. and Ringwood, J. V. (2003) First principles modelling of a pasteurisation plant for model predictive control. Mathematical and Computer Modelling of Dynamical Systems, 9(3), 281–301.
[68] Khadir, M. T., Richalet, J., Ringwood, J. V., and O'Connor, B. (2000a) Modelling and predictive control of milk pasteurisation in a Plate Heat Exchanger, in Proc. Foodsim, Nante, France, pp. 216–220.
[69] Khadir, M. T., Luo, G., and Ringwood, J. V. (2000b) Comparison of model predictive and PID controllers for dairy systems, in Proc. ISSC, Dublin, Rep of Ireland, pp. 442–449.
[70] Alpha Laval, Dairy Processing Handbook. Tetra Pak, 1995.
[71] Cybenko, G. (1989) Approximation by superpositions of a sigmoidal function, MCSS. Mathematics of Control, Signals and Systems 2, 303–314.
[72] Le Cun, Y., Denker, J. S., and Solla, S. A. (1989) Optimal brain damage, in Addvanced in Neural Information, vol 2, pp. 126–142.
[73] Nørgaard, M., Poulsen, N. K., and Hansen, L. K. (2000) Neural Networks for Modelling and Control of Dynamic Systems. London, UK: Springer-Verlag.
[74] Sjöberg, J. (1995) Non-Linear System Identification With Neural Networks. Ph.D. Thesis No 831, Division of Automatic Control, Department of Electrical Engineering, Linkping University, Sweden.
[75] Artificial Neural Network Toolbox for use with Matlab.
[76] Heicht-Nielsen, R. (1989) Neurocomputing. Reading, Massachusetts: Addison-Wesley.
[77] NAG Foundation Toolbox for use with Matlab, Numerical Algorithms group Ltd, 1995.
[78] Rao, C. V., Wright, S. J. and Rawlings, J. B. (1998) Application of interior-point mehods to model predictive control. Journal of Optimisation Theory and Applications, 99(3), 723–757.
[79] Khadir, M. T. (2020) Comparison of advanced and model predictive control for plate heat exchangers: application to pasteurization temperature control, in Pekař, Libor (ed.), Advanced Analytic and Control Techniques for Thermal Systems with Heat Exchangers, Chapter 20, pp. 433–456.

Index

www.ingramcontent.com/pod-product-compliance
Lightning Source LLC
Chambersburg PA
CBHW081524220326
41598CB00036B/6325
* 9 7 8 3 1 1 0 6 4 5 9 4 1 *